U0397166

NATURAL HISTORY
UNIVERSAL LIBRARY

西方博物学大系

主编：江晓原

THE NATURAL HISTORY OF BRITISH SHELLS

不列颠贝类志

[英] 爱德华·多诺万 著

华东师范大学出版社

图书在版编目(CIP)数据

不列颠贝类志 = The natural history of British shells : 英文 / (英) 爱德华 · 多诺万 (Edward Donovan)著. — 上海 : 华东师范大学出版社, 2018
(寰宇文献)
ISBN 978-7-5675-7996-5

Ⅰ. ①不… Ⅱ. ①爱… Ⅲ. ①贝类-英国-英文 Ⅳ. ①Q959.215

中国版本图书馆CIP数据核字(2018)第158790号

不列颠贝类志
The natural history of British shells
(英) 爱德华 · 多诺万 (Edward Donovan)

特约策划　黄曙辉　徐　辰
责任编辑　庞　坚
特约编辑　许　倩
装帧设计　刘怡霖

出版发行　华东师范大学出版社
社　　址　上海市中山北路3663号　邮编 200062
网　　址　www.ecnupress.com.cn
电　　话　021-60821666　行政传真　021-62572105
客服电话　021-62865537
门市（邮购）电话　021-62869887
地　　址　上海市中山北路3663号华东师范大学校内先锋路口
网　　店　http://hdsdcbs.tmall.com/

印 刷 者　虎彩印艺股份有限公司
开　　本　787 × 1092　16开
印　　张　47.5
版　　次　2018年8月第1版
印　　次　2018年8月第1次
书　　号　ISBN 978-7-5675-7996-5
定　　价　1200.00元（精装全一册）

出 版 人　王　焰

（如发现本版图书有印订质量问题，请寄回本社客服中心调换或电话021-62865537联系）

总　目

《西方博物学大系》总序

江晓原

《西方博物学大系》收录博物学著作超过一百种，时间跨度为 15 世纪至 1919 年，作者分布于 16 个国家，写作语种有英语、法语、拉丁语、德语、弗莱芒语等，涉及对象包括植物、昆虫、软体动物、两栖动物、爬行动物、哺乳动物、鸟类和人类等，西方博物学史上的经典著作大备于此编。

中西方“博物”传统及观念之异同

今天中文里的“博物学“一词，学者们认为对应的英语词汇是Natural History，考其本义，在中国传统文化中并无现成对应词汇。在中国传统文化中原有“博物”一词，与“自然史”当然并不精确相同，甚至还有着相当大的区别，但是在“搜集自然界的物品”这种最原始的意义上，两者确实也大有相通之处，故以“博物学”对译 Natural History 一词，大体仍属可取，而且已被广泛接受。

已故科学史前辈刘祖慰教授尝言：古代中国人处理知识，如开中药铺，有数十上百小抽屉，将百药分门别类放入其中，即心安矣。刘教授言此，其辞若有憾焉——认为中国人不致力于寻求世界“所以然之理”，故不如西方之分析传统优越。然而古代中国人这种处理知识的风格，正与西方的博物学相通。

与此相对，西方的分析传统致力于探求各种现象和物体之间的相互关系，试图以此解释宇宙运行的原因。自古希腊开始，西方哲人即孜孜不倦建构各种几何模型，欲用以说明宇宙如何运行，其中最典型的代表，即为托勒密（Ptolemy）的宇宙体系。

比较两者，差别即在于：古代中国人主要关心外部世界“如何”运行，而以希腊为源头的西方知识传统（西方并非没有别的知识传统，只是未能光大而已）更关心世界“为何”如此运行。在线性发展无限进步的科学主义观念体系中，我们习惯于认为“为何”是在解决了“如何”之后的更高境界，故西方的分析传统比中国的传统更高明。

然而考之古代实际情形，如此简单的优劣结论未必能够成立。例如以天文学言之，古代东西方世界天文学的终极问题是共同的：给定任意地点和时刻，计算出太阳、月亮和五大行星（七政）的位置。古代中国人虽不致力于建立几何模型去解释七政“为何”如此运行，但他们用抽象的周期叠加（古代巴比伦也使用类似方法），同样能在足够高的精度上计算并预报任意给定地点和时刻的七政位置。而通过持续观察天象变化以统计、收集各种天象周期，同样可视之为富有博物学色彩的活动。

还有一点需要注意：虽然我们已经接受了用“博物学”来对译 Natural History，但中国的博物传统，确实和西方的博物学有一个重大差别——即中国的博物传统是可以容纳怪力乱神的，而西方的博物学基本上没有怪力乱神的位置。

古代中国人的博物传统不限于“多识于鸟兽草木之名”。体现此种传统的典型著作，首推晋代张华《博物志》一书。书名“博物”，其义尽显。此书从内容到分类，无不充分体现它作为中国博物传统的代表资格。

《博物志》中内容，大致可分为五类：一、山川地理知识；二、奇禽异兽描述；三、古代神话材料；四、历史人物传说；五、神仙方伎故事。这五大类，完全符合中国文化中的博物传统，深合中国古代博物传统之旨。第一类，其中涉及宇宙学说，甚至还有“地动”思想，故为科学史家所重视。第二类，其中甚至出现了中国古代长期流传的“守宫砂”传说的早期文献：相传守宫砂点在处女胳膊上，永不褪色，只有性交之后才会自动消失。第三类，古代神话传说，其中甚至包括可猜想为现代“连体人”的记载。第四类，各种著名历史人物，比如三位著名刺客的传说，此三名刺客及所刺对象，历史上皆实有其人。第五类，包括各种古代方术传说，比如中国古代房中养生学说，房中术史上的传说人物之一“青牛道士封君达”等等。前两类与西方的博物学较为接近，但每一类都会带怪力乱神色彩。

“所有的科学不是物理学就是集邮”

在许多人心目中，画画花草图案，做做昆虫标本，拍拍植物照片，这类博物学活动，和精密的数理科学，比如天文学、物理学等等，那是无法同日而语的。博物学显得那么的初级、简单，甚至幼稚。这种观念，实际上是将“数理程度”作为唯一的标尺，用来衡量一切知识。但凡能够使用数学工具来描述的，或能够进行物理实验的，那就是“硬”科学。使用的数学工具越高深越复杂，似乎就越“硬”；物理实验设备越庞大，花费的金钱越多，似乎就越“高端”、越“先进”……

这样的观念，当然带着浓厚的“物理学沙文主义”色彩，在很多情况下是不正确的。而实际上，即使我们暂且同意上述“物理学沙文主义”的观念，博物学的“科学地位”也仍然可以保住。作为一个学天体物理专业出身，因而经常徜徉在“物理学沙文主义”幻影之下的人，我很乐意指出这样一个事实：现代天文学家们的研究工作中，仍然有绘制星图，编制星表，以及为此进行的巡天观测等等活动，这些活动和博物学家“寻花问柳”，绘制植物或昆虫图谱，本质上是完全一致的。

这里我们不妨重温物理学家卢瑟福(Ernest Rutherford)的金句：“所有的科学不是物理学就是集邮（All science is either physics or stamp collecting）。”卢瑟福的这个金句堪称“物理学沙文主义”的极致，连天文学也没被他放在眼里。不过，按照中国传统的“博物”理念，集邮毫无疑问应该是博物学的一部分——尽管古代并没有邮票。卢瑟福的金句也可以从另一个角度来解读：既然在卢瑟福眼里天文学和博物学都只是“集邮”，那岂不就可以将博物学和天文学相提并论了？

如果我们摆脱了科学主义的语境，则西方模式的优越性将进一步被消解。例如，按照霍金（Stephen Hawking）在《大设计》（*The Grand Design*）中的意见，他所认同的是一种“依赖模型的实在论（model-dependent realism）”，即“不存在与图像或理论无关的实在性概念（There is no picture- or theory-independent concept of reality）”。在这样的认识中，我们以前所坚信的外部世界的客观性，已经不复存在。既然几何模型只不过是对外部世界图像的人为建构，则古代中国人干脆放弃这种建构直奔应用（毕竟在实际应用中我们只需要知道七政“如何”运行），又有何不可？

传说中的“神农尝百草”故事，也可以在类似意义下得到新的解读：“尝百草”当然是富有博物学色彩的活动，神农通过这一活动，得知哪些草能够治病，哪些不能，然而在这个传说中，神农显然没有致力于解释“为何”某些草能够治病而另一些则不能，更不会去建立“模型”以说明之。

“帝国科学”的原罪

今日学者有倡言“博物学复兴”者，用意可有多种，诸如缓解压力、亲近自然、保护环境、绿色生活、可持续发展、科学主义解毒剂等等，皆属美善。编印《西方博物学大系》也是意欲为“博物学复兴”添一助力。

然而，对于这些博物学著作，有一点似乎从未见学者指出过，而鄙意以为，当我们披阅把玩欣赏这些著作时，意识到这一点是必须的。

这百余种著作的时间跨度为 15 世纪至 1919 年，注意这个时间跨度，正是西方列强“帝国科学”大行其道的时代。遥想当年，帝国的科学家们乘上帝国的军舰——达尔文在皇家海军“小猎犬号”上就是这样的场景之一，前往那些已经成为帝国的殖民地或还未成为殖民地的“未开化”的遥远地方，通常都是踌躇满志、充满优越感的。

作为一个典型的例子，英国学者法拉在（Patricia Fara）《性、植物学与帝国：林奈与班克斯》（*Sex, Botany and Empire, The Story of Carl Linnaeus and Joseph Banks*）一书中讲述了英国植物学家班克斯（Joseph Banks）的故事。1768 年 8 月 15 日，班克斯告别未婚妻，登上了澳大利亚军舰“奋进号”。此次“奋进号”的远航是受英国海军部和皇家学会资助，目的是前往南太平洋的塔希提岛（Tahiti，法属海外自治领，另一个常见的译名是“大溪地”）观测一次比较罕见的金星凌日。舰长库克（James Cook）是西方殖民史上最著名的舰长之一，多次远航探险，开拓海外殖民地。他还被认为是澳大利亚和夏威夷群岛的“发现”者，如今以他命名的群岛、海峡、山峰等不胜枚举。

当“奋进号”停靠塔希提岛时，班克斯一下就被当地美丽的土著女性迷昏了，他在她们的温柔乡里纵情狂欢，连库克舰长都看不下去了，“道德愤怒情绪偷偷溜进了他的日志当中，他发现自己根本不可能不去批评所见到的滥交行为”，而班克斯纵欲到了“连嫖妓都毫无激情”的地步——这是别人讽刺班克斯的说法，因为对于那时常年航行于茫茫大海上的男性来说，上岸嫖妓通常是一项能够唤起“激情”的活动。

而在“帝国科学”的宏大叙事中，科学家的私德是无关紧要的，人们关注的是科学家做出的科学发现。所以，尽管一面是班克斯在塔希提岛纵欲滥交，一面是他留在故乡的未婚妻正泪眼婆娑地“为远去的心上人绣织背心”，这样典型的“渣男”行径要是放在今天，非被互联网上的口水淹死不可，但是“班克斯很快从他们的分离之苦中走了出来，在外近三年，他活得倒十分滋润”。

法拉不无讽刺地指出了“帝国科学”的实质：“班克斯接管了当地的女性和植物，而库克则保护了大英帝国在太平洋上的殖民地。”甚至对班克斯的植物学本身也调侃了一番：“即使是植物学方面的科学术语也充满了性指涉。……这个体系主要依靠花朵之中雌雄生殖器官的数量来进行分类。”据说“要保护年轻妇女不受植物学教育的浸染，他们严令禁止各种各样的植物采集探险活动。”这简直就是将植物学看成一种“涉黄”的淫秽色情活动了。

在意识形态强烈影响着我们学术话语的时代，上面的故事通常是这样被描述的：库克舰长的“奋进号”军舰对殖民地和尚未成为殖民地的那些地方的所谓“访问”，其实是殖民者耀武扬威的侵略，搭载着达尔文的“小猎犬号”军舰也是同样行径；班克斯和当地女性的纵欲狂欢，当然是殖民者对土著妇女令人发指的蹂躏；即使是他采集当地植物标本的“科学考察”，也可以视为殖民者“窃取当地经济情报”的罪恶行为。

后来改革开放，上面那种意识形态话语被抛弃了，但似乎又走向了另一个极端，完全忘记或有意回避殖民者和帝国主义这个层面，只歌颂这些军舰上的科学家的伟大发现和成就，例如达尔文随着“小猎犬号”的航行，早已成为一曲祥和优美的科学颂歌。

其实达尔文也未能免俗，他在远航中也乐意与土著女性打打交道，当然他没有像班克斯那样滥情纵欲。在达尔文为“小猎犬号”远航写的《环球游记》中，我们读到：“回程途中我们遇到一群黑人姑娘在聚会，……我们笑着看了很久，还给了她们一些钱，这着实令她们欣喜一番，拿着钱尖声大笑起来，很远还能听到那愉悦的笑声。”

有趣的是，在班克斯在塔希提岛纵欲六十多年后，达尔文随着“小猎犬号”也来到了塔希提岛，岛上的土著女性同样引起了达尔文的注意，在《环球游记》中他写道：“我对这里妇女的外貌感到有些失望，然而她们却很爱美，把一朵白花或者红花戴在脑后的髮髻上……”接着他以居高临下的笔调描述了当地女性的几种发饰。

用今天的眼光来看，这些在别的民族土地上采集植物动物标本、测量地质水文数据等等的“科学考察”行为，有没有合法性问题？有没有侵犯主权的问题？这些行为得到当地人的同意了吗？当地人知道这些行为的性质和意义吗？他们有知情权吗？……这些问题，在今天的国际交往中，确实都是存在的。

也许有人会为这些帝国科学家辩解说：那时当地土著尚在未开化或半开化状态中，他们哪有“国家主权”的意识啊？他们也没有制止帝国科学家的考察活动啊？但是，这样的辩解是无法成立的。

姑不论当地土著当时究竟有没有试图制止帝国科学家的“科学考察”行为，现在早已不得而知，只要殖民者没有记录下来，我们通常就无法知道。况且殖民者有军舰有枪炮，土著就是想制止也无能为力。正如法拉所描述的：“在几个塔希提人被杀之后，一套行之有效的易货贸易体制建立了起来。”

即使土著因为无知而没有制止帝国科学家的“科学考察”行为，这事也很像一个成年人闯进别人的家，难道因为那家只有不懂事的小孩子，闯入者就可以随便打探那家的隐私、拿走那家的东西、甚至将那家的房屋土地据为己有吗？事实上，很多情况下殖民者就是这样干的。所以，所谓的“帝国科学”，其实是有着原罪的。

如果沿用上述比喻，现在的局面是，家家户户都不会只有不懂事的孩子了，所以任何外来者要想进行“科学探索”，他也得和这家主人达成共识，得到这家主人的允许才能够进行。即使这种共识的达成依赖于利益的交换，至少也不能单方面强加于人。

博物学在今日中国

博物学在今日中国之复兴，北京大学刘华杰教授提倡之功殊不可没。自刘教授大力提倡之后，各界人士纷纷跟进，仿佛昔日蔡锷在云南起兵反袁之“滇黔首义，薄海同钦，一檄遥传，景从恐后”光景，这当然是和博物学本身特点密切相关的。

无论在西方还是在中国，无论在过去还是在当下，为何博物学在它繁荣时尚的阶段，就会应者云集？深究起来，恐怕和博物学本身的特点有关。博物学没有复杂的理论结构，它的专业训练也相对容易，至少没有天文学、物理学那样的数理“门槛”，所以和一些数理学科相比，博物学可以有更多的自学成才者。这次编印的《西方博物学大系》，卷帙浩繁，蔚为大观，同样说明了这一点。

最后，还有一点明显的差别必须在此处强调指出：用刘华杰教授喜欢的术语来说，《西方博物学大系》所收入的百余种著作，绝大部分属于“一阶”性质的工作，即直接对博物学作出了贡献的著作。事实上，这也是它们被收入《西方博物学大系》的主要理由之一。而在中国国内目前已经相当热的博物学时尚潮流中，绝大部分已经出版的书籍，不是属于“二阶”性质（比如介绍西方的博物学成就），就是文学性的吟风咏月野草闲花。

要寻找中国当代学者在博物学方面的“一阶”著作，如果有之，以笔者之孤陋寡闻，唯有刘华杰教授的《檀岛花事——夏威夷植物日记》三卷，可以当之。这是刘教授在夏威夷群岛实地考察当地植物的成果，不仅属于直接对博物学作出贡献之作，而且至少在形式上将昔日“帝国科学”的逻辑反其道而用之，岂不快哉！

2018 年 6 月 5 日

于上海交通大学

科学史与科学文化研究院

出版说明

《不列颠贝类志》是爱德华·多诺万（Edward Donovan，1768—1837）的一部博物学著作。多诺万是英国博物学家，生于爱尔兰的科克，曾是伦敦林奈学会会员。在那个博物学家和收藏家纷纷开设个人博物馆的时代，1807 年他也在伦敦设立了博物学研究所，展示数百件动物标本与植物标本。他本人并不去海外采集标本，而是运用自己良好的社会关系，委托包括约瑟夫·班克斯和詹姆斯·库克在内的诸多探险家为自己工作，因而得以聚集起大量罕见标本。对这些标本的研究帮助他出版了诸多负有盛名的博物学著作，如《不列颠珍稀鸟类志》《中国昆虫志》《印度昆虫志》。多诺万亲自制作铜版并调色，使书中的画色彩尽可能鲜艳逼真，加上展示了一批与他有特殊关系的相关人士搞到的珍稀标本，因此这些著作一经出版其声名就居于当时同类作品的最前列。

《不列颠贝类志》初版刊行于 1779 年，是多诺万的成名作之一。本书将眼光放在当时博物学研究并不太重视的贝类身上，对不列颠岛周边生息的双壳纲、腹足纲和掘足纲生物进行了分类描述。全书凡五卷，共 1200 页，配有约 150 幅根据实体标本测量、描绘的精美彩图。在绘制这些插画时，多诺万不仅进行贝类的单体展示，而且常常根据实际分类通过巧妙的构图，将若干大小各异的标本放在一张图中，显示其实际大小比例，令画面尤为生动。

今据原版影印。

THE

NATURAL HISTORY

OF

BRITISH SHELLS,

INCLUDING

FIGURES AND DESCRIPTIONS

OF ALL THE

SPECIES HITHERTO DISCOVERED IN GREAT BRITAIN,

SYSTEMATICALLY ARRANGED

IN THE LINNEAN MANNER,

WITH

SCIENTIFIC AND GENERAL OBSERVATIONS ON EACH.

VOL. I.

By E. DONOVAN, F.L.S.

AUTHOR OF THE NATURAL HISTORIES OF

BRITISH BIRDS, INSECTS, &c. &c.

LONDON

PRINTED FOR THE AUTHOR,

AND FOR

F. AND C. RIVINGTON, No 62, ST. PAUL'S CHURCH-YARD.

1799.

THE

NATURAL HISTORY

OF

BRITISH SHELLS.

INTRODUCTION.

VERMES.

THIS class of Animals was formerly confounded with Insects and Plants: the *Intestina* and *Mollusca* were referred to the first class: the *Zoophyta* and *Lithophyta* to the latter; and some Authors had even classed the *testacea*, or Shells, as a branch of Mineralogy, without regarding the Animals inhabiting them. Linnæus, in the *Systema Naturæ*, comprehends the whole of these creatures in the last class of Zoology; and forms their classical character from their internal structure, as in larger and more perfect animals: COR uniloculare, inauritum; *Sanie* frigida, albida. Tentaculatis *Vermibus*. Heart furnished with one ventricle, without auricle; *sanies* cold and whitish, or colourless. The five orders of the Linnæan class *Vermes* are thus defined:—

Intestina, simple, naked, destitute of limbs.

Mollusca, simple, naked; but not without limbs.

Testacea, animal with a calcareous covering.

Lithophyta, animal composite, affixed to, and fabricate a calcareous base.—Coral.

Zoophyta, a vegetating stem like a plant; animal composite, and resemble flowers.

Linnæus has included in the *Testacea* Order the whole tribe of Shells. In the generic characters he regards both the Shell and its inhabitant: in the definition of species, the former only is attended to. There are very strong arguments against the method of arranging this tribe by the Animals, although it cannot be denied, that the Shells are only the coverings or habitations, and should not demand our primary attention *.

The TESTACEA are Vermes of the soft and simple kind, and are covered with a calcareous habitation. These are separated into three divisions, according to the number of valves of which the Shell consists. The first division includes only three genera, *Chiton*, *Lepas*, and *Pholas*; these are called Multivalves, and are formed of many valves, or pieces, disposed transversely on each other. The second division consists of Bivalves, or Shells of two pieces, connected together with a hinge, or cartilage. The third division is of Univalves, and have the Shell complete in one piece, as the word implies. The Linnæan genera are—

* Vide DONOVAN's Instructions for collecting and preserving Subjects of Natural History. London, 1794.

Multivalvia.

CHITON.	LEPAS.	PHOLAS.

Bivalvia : conchæ.

MYA.	SOLEN.	TELLINA.
CARDIUM.	MACTRA.	DONAX.
VENUS.	SPONDYLUS.	CHAMA.
ARCA.	OSTREA.	ANOMIA.
MYTILLUS.	PINNA.	

Univalvia.

spira regulari COCHLEAE.

ARGONAUTA.	NAUTILLUS.	CONUS.
CYPRAEA.	BULLA.	VOLUTA.
BUCCINUM.	STROMBUS.	MUREX.
TROCHUS.	TURBO.	HELIX.
NERITA.	HALIOTIS.	

sine spira regulari.

PATELLA.	DENTALIUM.	SERPULA.
TEREDO.	SABELLA.	

Pecten

1

1

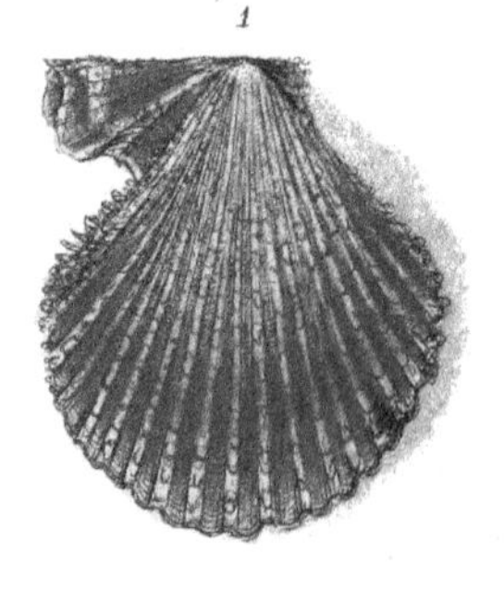

1

2

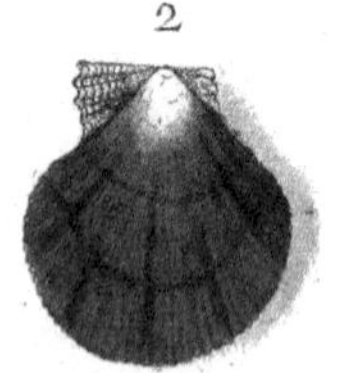

1

PLATE I.

FIG. I. I. I. I.

OSTREA VARIA.

VARIEGATED, OR ONE-EARED SCALLOP.

GENERIC CHARACTER.

Animal a Tethys. Shell bivalve unequal. The hinge without a tooth, having a small oval cavity.

SPECIFIC CHARACTER
AND
SYNONYMS.

Shell almost equally convex; about thirty rays, scabrous, imbricated, or beset with transverse scales. One ear *.

OSTREA VARIA testa æquivalvi: radiis triginta scabris compressis echinatis uni aurita. *Gmel.—Linn. Syst. Nat.* 3324. 48.

P. subrufus, striis viginti quatuor, ad minimum donatus.—P. parvus, ex croceo variegatus, tenuiter admodum striatus, alternis fere striis paulo minoribus. *List. H. Conch.*

Pecten minor nostras, striis plurimis minoribus. *Mus. Petiv. p.* 86. *No.* 830.

Pectunculus echinatus fusco purpureus. *Borlase Corn. p.* 277.

* It has two ears, but one is considerably larger than the other.

PLATE I.

Pecten varius : variegated scallop. *Pen. Br. Zool. No.* 64. *tab.* 61. *fig.* 64.

PECTEN MONOTIS : ONE EARED ESCALLOP. Parvus angustior, æquivalvis, inæqualiter auritus, strigis echinatis. *Da Costa. Tab.* 10. *fig.* 1. 2. 4. 5. 7. 9.

Many beautiful kinds of this species are found on our coasts. Some are of an uniform, obscure, reddish, or purple colour, without any markings : some are violet, and others bright yellow, or orange. The most elegant kinds are variegated with different colours, as white, red, purple, and brown. The purple kind marbled with irregular spots, and waves of white ; and the coral red, with black and white markings, and white on the upper part, are select specimens of these elegant varieties.

Pennant says, this species is often found in oyster-beds, and dragged up with them. " It is frequent on most of the shores of England ; as in Wales ; at Margate, and Sheerness, in Kent ; in Sussex and Dorsetshire ; in Devonshire ; at Lelant and Whitsand Bay, &c. in Cornwall ; the ostium of the river Aln in Northumberland, and many other places." *Da Costa.*

PLATE I.

FIG. II.

PECTEN OBSOLETUS.

GENERIC CHARACTER.

Ostrea. *Linn.*

SPECIFIC CHARACTER AND SYNONYMS.

One large striated ear, with smooth equal shells; eight *obsolete* rays; of a dark purple colour. *Penn. Br. Zool. No.* 66. *tab.* 61. *fig.* 66.

PECTEN PARVUS: parvus fuscus longitudinaliter striatus, *Da Costa. Br. Conch.* 153. 8.

This is a very rare species; da Costa received his specimen from Cornwall.

The valves are equal and shallow; the shell thin, and semitransparent; the ears unequal, one being very small. The inside is smooth and brown, with a pearly gloss. The outside is a dull purplish brown, with numerous fine longitudinal striæ *, eight or ten of which are more prominent than the rest. These are surely not the *obsolete* rays of Pennant, as da Costa imagines; the former author must allude to the intermediate rays which are depressed and appear worn, as he describes them.

* The figure in Pennant's work is represented with transverse striæ; this appears however, to be an error of the engraver.

2

PLATE II.

FIG. I. I.

TURBO CIMEX.

LATTICED WHELKE.

GENERIC CHARACTER.

Animal Limax. Univalve, spiral, or of a taper form. Aperture somewhat compressed, orbicular, entire.

SPECIFIC CHARACTER

AND

SYNONYMS.

Shell oblong-oval. Striæ decussate, or intersect each other in a spiral direction.

TURBO CIMEX, testa oblongo-ovata, striis decussatis: punctis eminentibus. *Lin. Syst. Nat. p.* 1233. *No.* 609.

Turbo Cancellatus, *Latticed*. Turbo minimus albus cancellatim vel decussatim striatus. *Da Costa Br. Conch.* 104. 60. *tab.* 8. *fig.* 6. 9.

The natural size of this shell is shewn at Fig. I. together with its microscopic appearance. It is a very small species, thick, without gloss. The striæ are elevated, broad, and cross each other so as to form a deep latticed-work of thick ridges. This species is noted from Cornwall and Guernsey: it is also found in the Mediterranean.

FIG. II. III. IV. VI. V.

TURBO PULLUS.

PAINTED WHELKE.

GENERIC CHARACTER.

Animal Limax. Shell univalve, spiral, or of a taper form. Aperture rather compressed, orbicular, entire.

SPECIFIC CHARACTER AND SYNONYMS.

TURBO PULLUS. Turbo testa imperforata ovata lævi, apertura antice diducta. *Linn. Syst. Nat. p.* 1233. *No.* 610.

Turbo minimus lævis, variegatus, albo rubicundus. Small red and white variegated Whelke. *Borlase Cornw. p.* 277.

Painted, Turbo pictus. Turbo minimus lævis, albo et rubro perbelle pictus, *da Costa, p.* 103. 59. *tab.* 8. *fig.* 1. 3.

A minute, but elegant species; it is a very delicate shell, thin and transparent, smooth and glossy. The varieties are numerous; generally white or blush-rose colour, with the markings crimson or reddish purple, disposed in zones, spiral circles, transverse streaks, irregular waves, lines, spots, and specklings. Some are variegated with different shades of brown in a similar manner.

Fig. II. represents the natural size: Fig. III. a full grown specimen. Fig. IV. IV. IV. are elegant varieties, as they appear under the microscope. Da Costa notes this species from the coast of Cornwall, and from Exmouth in Devonshire.

3

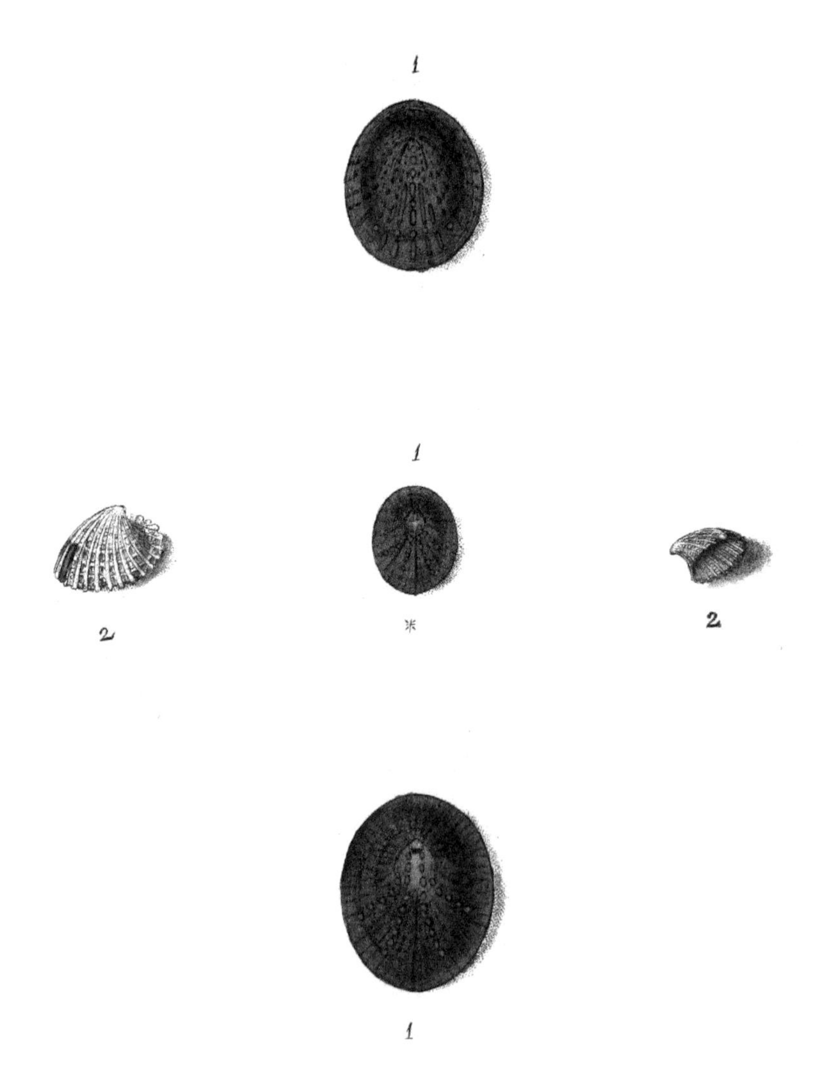

PLATE III.

FIG. I*. I. I.

PATELLA PELLUCIDA.

BLUE RAYED LIMPET.

GENERIC CHARACTER.

Animal Limax. Shell univalve, subconic, without spires.

SPECIFIC CHARACTER AND SYNONYMS.

Shell entire, gibbous, pellucid, marked with longitudinal blue rays.

PATELLA PELLUCIDA: testa integerrima obovato gibba pellucida: radiis quatuor cæruleis. *Gmelin. Lin. Syst. Nat.* 3717. 153.

Patella minima lævis pellucida, aliquot cæruleis lineis eleganter insignita. *List. Hist. Conch. tab.* 543. *fig.* 27.

Patella minor, fusca, tenuis, umbone nigro ad extremitatem anteriorem detruso, tribus inde lineis cæruleis per dorsum decurrentibus pulchre distincta. *Wallace, Orkneys, p.* 41.

Patella Anglica parva, prætenuis cymbuliformis, lineis cæruleis guttatis. *Mus. Petiv. cent.* 8. *p.* 68. *No.* 725.

PLATE III.

Transparent Patella. *Br. Zool.* 4. *No.* 150. *tab.* 90. *fig.* 150.

Patella Lævis. Smooth Patella. *Br. Zool. No.* 151. an *old shell.*

Patella Cæruleata. Blue rayed. *Da Costa. Br. Conch.* 7. 4. *tab.* 1. *fig.* 5. 6.

Lepas d'eau douce demi-ovoide transparent, a trois lignes bleues. *D'Avila, tab.* 1. *p.* 428. *No.* 962.

In the young state, this shell is very transparent and horny, the aperture ovoid, and the margins smooth and level; it has also several longitudinal lines of bright blue colour, which extend from the vertex down the back to the margin. According to Linnæus, these should be four in number; some authors say five, and *Borlase* mentions nine. The blue colour is disposed in spots in some specimens; in others in lines; and again in some others in short and interrupted dashes. Linnæus observes that the bright blue colour has not been found in any Shell except this.

The old shells are very different from the young ones, and have been mistaken by some Authors for distinct species. The young shell is remarkable for its pellucidity. The old ones are thicker and larger: the aperture irregular: the vertex two-thirds of the shell; and the rays of blue, dusky. Fig. I. * represents the natural size.

The Shell is found on the coast of Cornwall, and on the Dorset coast, near Weymouth. *Martin, Sibbald,* and *Wallace,* received it from the western isles of Scotland and the Orkneys.

PLATE III.

PATELLA FISSURA.

SLIT-LIMPET.

GENERIC CHARACTER.

Animal limax. Shell univalve, subconic, without spires.

SPECIFIC CHARACTER
AND
SYNONYMS.

Oval, striated, reticulated. Vertex recurved, or bent back. A slit in the anterior part.

Patella Fissura : testa ovali striato-reticulata : vertice recurvo, anterius fissa. *Gmelin, Linn. Syst. Nat.* 3728. 192.

Patella integra parva, alba, cancellata, fissura notabili in margine.
List. H. Conch. tab. 543.
Petiv. Gaz. tab. 75. *fig.* 2.

Patella testa sulcato-reticulata, vertice recurvo, margine antice sursum fisso. *Müller-zool-dan.* 1. *p.* 83. *t.* 24. *f.* 7. 9.
rar. 1. *p.* 51. *prodr.* 2864.

Patella fissura. Slit. *Br. Zool. t.* 90. *f.* 152. *p.* 144.
Da Costa Br. Conch. 11. 5. *tab.* 1. *fig.* 4.

Lepas d'eau douce reticulé, avec une petite fente, ou entaille.
D'Avila, Cab. 1. *p.* 428. *No.* 962.

Found on the coasts of Cornwall and Devonshire.

4

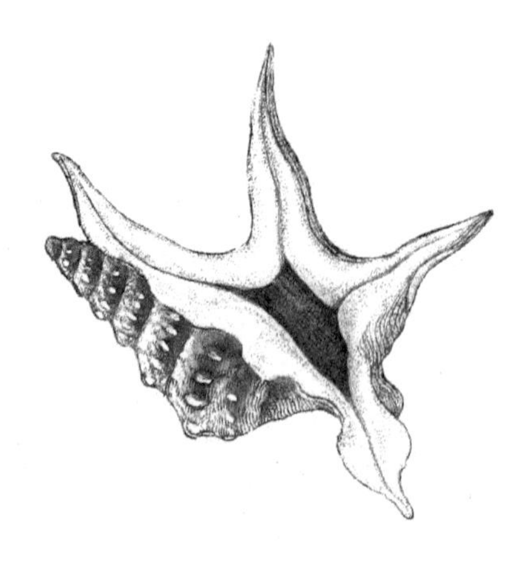

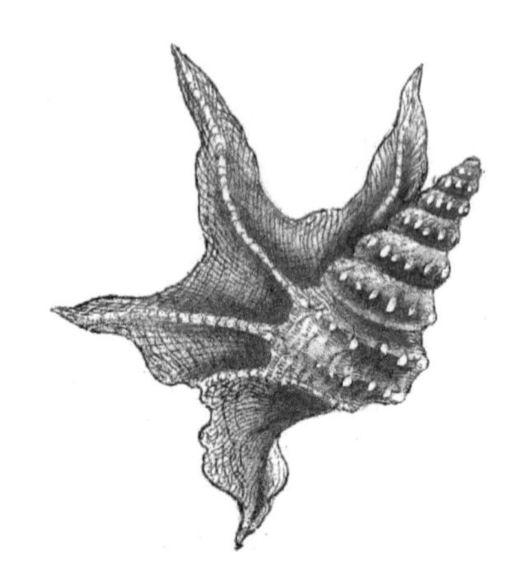

PLATE IV.

STROMBUS PES PELECANI.

CORVORANT'S FOOT.

GENERIC CHARACTER.

Animal a slug. Shell univalve, spiral. The aperture much dilated, and lip expanding into a groove.

SPECIFIC CHARACTER

AND

SYNONYMS.

Lip expanded, divided into four fingers or prongs.

STROMBUS PES PELECANI: testa labro tetradactylo palmato digitis angulato, fauce lævi. *Gmel.—Lin. Syst. Nat.* 3507. 2.

Cochlea testa longa acuminata, aperturæ labro dilatato, duplici stria antice sinuato. *Lin. Fn. Suec.* 1. *p.* 378. *No.* 1323.

Aporrhais Quadrifidus. Four-fingered. Aporrhais subfuscus, anfractibus nodosis, labro palmato quadrifido. *Da Costa Br. Conch.* 136. 80. *Tab.* 7. *fig.* 7.

Buccinum bilingue striatum labro propatulo digitato. *Lister H Conch. tab.* 8. 65. *fig.* 20.

Strombus canaliculatus, rostratus, ore labioso, striatus, papillosus, auritus aure admodum crassa, et in quatuor appendices breviores expansa, ex candida cinereus. *Gualt.* 1. *Conch. tab.* 53. *fig. A.*

PLATE IV.

Aporrhais Edinburgicus minor nodoso. *Petiv. Gaz. tab.* 79. *fig.* 6. —*tab.* 127. *fig.* 11.

Strombus Pes pelicani, Corvorant's foot. *Penn. Br. Zool. No.* 94. *tab.* 75. *fig.* 94.

Aile de Chauve Souris femelle, Patte D'Oye, ou Hallebarde. *D'Avila Cab. p.* 191. *No.* 344.

A very singular, but not uncommon shell on some of our coasts, as Cornwall, Devonshire, Durham and Sussex. In Carnarvonshire and Merionethshire, in Wales, on the coast of Scotland, and in the Orkneys.

5

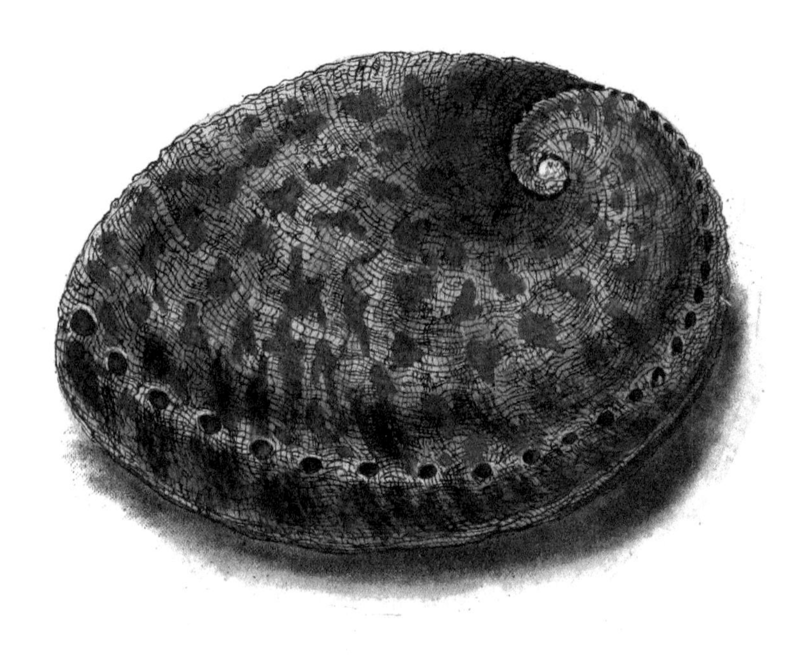

PLATE V.

HALIOTIS TUBERCULATA.

TUBERCULATED SEA EAR.

GENERIC CHARACTER.

Animal a slug. Shell univalve, dilated, or flat, almost open a row of orifices its length, spire near one end turned in.

SPECIFIC CHARACTER

AND

SYNONYMS.

Oblong-oval. Outside furrowed transversely, rugged, tuberculated.

HALIOTIS TUBERCULATA, testa subovata, dorso transversim rugoso tuberculato. *Gmel.—Linn. Syst. Nat. Conch. p.* 3687. *sp.* 2.

Auris marina, major profunde sulcata, magis depressa, fusco colore obsita, intus argentea. *Gualt. Ind. Conch. tab.* 69. *fig.* 1.

Auris marina quibusdam: Patelli fera Rondoletii, λεπὰς ἄγεια Aristotelis; Mother of Pearl, Anglice. *List. H. An. Angl. p.* 167. *tit.* 16. *tab.* 3. *fig.* 16.

Tuberculated Sea Ear. *Pennant Br. Zool. No.* 144. *tab.* 88. *fig.* 144.

Haliotis Vulgaris. Common Sea Ear. *Da Costa, Br. Conch. p.* 15. *pl.* 2. *fig.* 1, 2.

Pennant says this species is frequently cast upon the southern coast of Devonshire. It is common on the eastern coast of Sussex; and on the coast of the isle of Guernsey.—It adheres like limpets, to the rocks, when living.

6

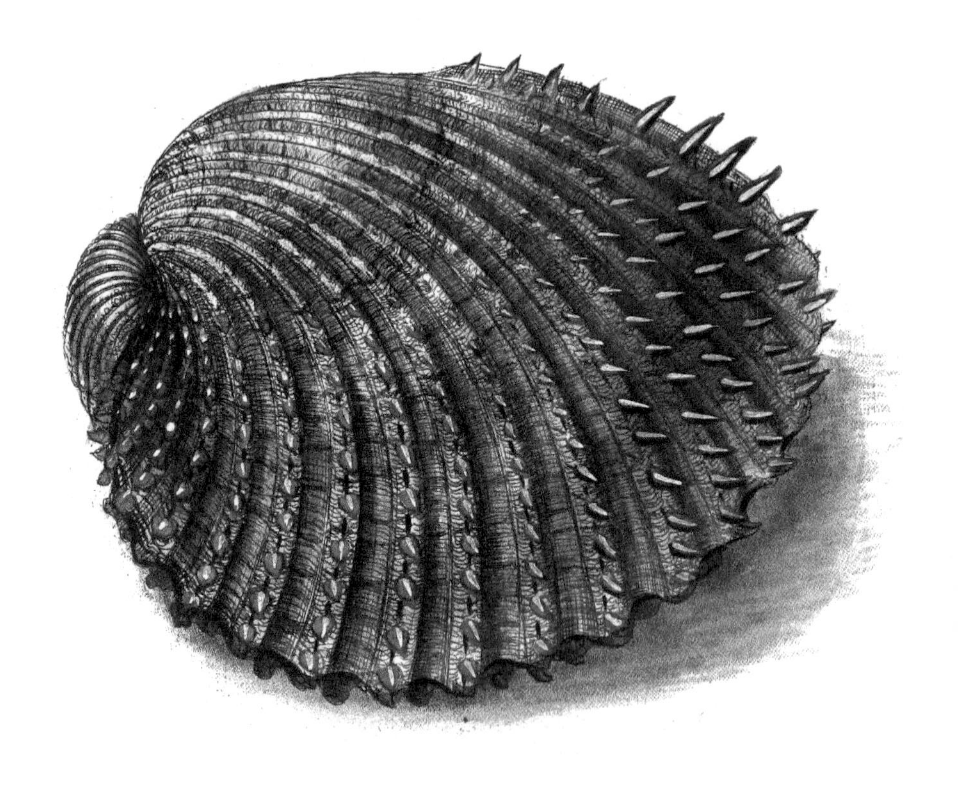

PLATE VI.

CARDIUM ACULEATUM.

SPIKED COCKLE.

GENERIC CHARACTER.

Two teeth near the beak; and another remote one on each side of the shell.

SPECIFIC CHARACTER

AND

SYNONYMS.

Shell nearly heart-shaped. Ribs high, sulcated down the middle, and beset with long canaliculated spines.

CARDIUM ACULEATUM: C. testa subcordata: sulcis convexis linea exaratis: exterius aculeato ciliatis.—*Gmel. Linn. Syst. Conch. p.* 3247. 7.

Pectunculus maximus insigniter echinatus. *Wallace, Orkney. p.* 44.

Cœur de bœuf. *Argenville. Conch. I. p.* 335. *fig. B.*

Cœur de bœuf Epineux. *D'Avilla Cab. p.* 355. *No.* 817.

Concha cordiformis æquilatera, umbone cardium unito, striata, striis latis canaliculatis muricata aculeis longis et acutis, aliquando recurvis in summitate striarum positis, albida, et parvis maculis luteis obscure fasciata. *Gualt. I. Conch. tab.* 72. *fig.* A.

Cardium Aculeatum. Aculeated. *Penn. Br. Zool.* 137. *tab.* 50. *fig.* 37.

Cardium Aculeatum, Spiked Cockle, A. *Da Costa, Br. Conch. p.* 175.

D 2

PLATE VI.

This is the largest of the Cardium, or Cockle genus, that inhabits any of the British shores. It is noted by Wallace as a Shell of the Orkneys; and also by Pennant, who likewise found it off the Hebrides.—It is a thick shell, with high radiated ribs, and beset with large processes or spines that are hollowed. It is covered with a fibrous epidermis, of a blackish colour, varied with light browns; the colour beneath is white, with a faint tint of red, or rose colour.—Marginal circumference ten inches and an half.

This is one of the six rare species Da Costa could not procure for his work, and to which he alludes in the preface*.

* I have described the shells from the objects themselves, except in *six instances*, where I could not procure the originals to complete the series; in which case I have borrowed them from authors of veracity; and the *Reader* will find those species distinguished by Roman characters.

PLATE VII.

LEPAS ANATIFERA.

ANATIFEROUS ACORN SHELL.

GENERIC CHARACTER.

Animal. triton. Shell of many unequal valves; affixed by a stem.

SPECIFIC CHARACTER
AND
SYNONYMS.

Shell compressed, consists of five parts, affixed to a pedicle, or membraneous tube.

LEPAS ANATIFERA, testa compressa quinquevalvi lævi pedunculo insidente. *Gmel. Linn. Syst. Nat. Conch. p.* 3211. 13.

Balanus Anatifera compressus quinquevalvis lævis, tubo seu colle membranaceo insidente. *Da Costa, Br. Conch.* 253. 72. *tab.* 18. *fig.* 3.

Concha quinquevalvis compressa, tubulo quodam lignis aut algæ marinæ adhærens; animal sui generis multis cirrhis instructum continens, falso dicta anatifera. *Sibbald. Mus. p.* 170. *No* 2.

Lepas testa compressa basi membrana cylindricea. *Fn. Suec. I. n.* 1350.

Lepas Anatifera cum Tritone. *Stalp. Obs.* 2, *p.* 458. *t.* 15. *Osb. it.* 82.

D 3

PLATE VII.

Barnacle Shell, or Concha anatifera. *Merret. Pin. p.* 194.
Balanus Compressa, Flat centre Shell. *Grew. Mus. p.* 148.
Wallace, Orkn. p. 45. *fig.* 1.—*Mus. Petiv. p.* 82. *No.* 802.
Anatiferous. *Br. Zool. No.* 9. *tab.* 38. *fig.* 9.
Concha anatifera marfine læve. *List. H. Conch. tab.* 440. *fig.* 283.
β Concha anatifera subrotunda Bartholini. *Lister. Conch. t.* 439. *f.* 280.
γ Tellina cancellifera striis minimis argutissime signata cinerea. *Gualt. testac. t.* 106. *f. B.*

The Lepas Anatifera is found on the coasts of England and Ireland, but more frequently on that of Scotland. It adheres by means of its branches, or pedicles, to the bottoms of ships, planks, logs, and other substances floating in the water.

This curious marine production consists of many unequal membraneous branches, or arms, at the ends of which the Shells are disposed in an irregular manner; the larger clustering with the smaller in groups, and forming bunches of various sizes. The branches are of a fine red; the Shells of a bluish violet. The animal within is a *triton*, and is furnished with many *cirrhi*, or *tentacula*, with which it takes its food. These tentacula are pectinated like feathers, and hang out of the Shells when open. In the sixteenth century they were, in fact, supposed to be feathers, and hence arose the whimsical belief that a barnacle produced a *goose**. Nor was this a vulgar opinion only; it was sanctioned by the grave details of learned naturalists of that time,

* *Vide* Anas Albifrons, *Plate* 102.—*Hist. Brit. Birds.*

and particularly by *Gerard**, whose observations are generally noticed by authors, in describing this curious species.

* "What our eyes have seene, and hands have touched, we shall declare. There is a small island in *Lancashire* called the *Pile* of *Foulders*, wherein are found the broken pieces of old and bruised ships, some whereof have been cast thither by shipwrake, and also the trunks and bodies with the branches of old and rotten trees, cast up there likewise; whereon is found a certaine spume, or froth, that in time breedeth unto certaine shels, in shape like those of the muskle, but sharper pointed, and of a whitish colour, wherin is contained a thing in form like a lace of silke finely woven, as it were, together, of a whitish colour; one end whereof is fastened unto the inside of the shell, even as the fish of oisters and muskles are: the other end is made fast unto the belly of a rude masse, or lumpe, which in time commeth to the shape and form of a bird. When it is perfectly formed, the shell gapeth open, and the first thing that appeareth is the foresaid lace or string; next come the legs of the bird, hanging out, and as it groweth greater it openeth the shell by degrees, till at length it is all come forth, and hangeth onely by the bill: in short space after it commeth to full maturitie, and falleth into the sea, where it gathereth feathers, and groweth to fowle bigger than a Mallard and lesser than a Goose, having blacke legs and bill or beake, and feathers blacke and white, spotted in such manner as is our *Magpie*, called in some places a Pie-Annet, which the people of *Lancashire* call by no other name than a *tree Goose*: which place aforesaid, and all those parts adjoyning, do so much abound therewith, that one of the best is bought for three-pence. For the truth hereof, if any doubt, may it please them to repaire unto me, and I shall satisfie them by the testimonie of good witnesses." Vide GERARD'S HERBAL, p. 1587, 1588.

8

PLATE VIII.

FIG. I.

TROCHUS MAGUS.

TUBERCULATED TOP SHELL.

GENERIC CHARACTER.

Animal a slug. Shell conic. Aperture nearly triangular.

SPECIFIC CHARACTER AND SYNONYMS.

Pyramidal somewhat depressed; base umbilicated. The ridges of the spires rising into distinct tubercles.

TROCHUS MAGUS, testa oblique umbillicata: convexa, anfractibus supra obtuse nodulosis.—*Gmel. Linn. Syst. Nat. Conch. p.* 3567.—*Sp.* 7.

Trochus acuminatus, crebris striis transverse et undatim dispositis donatus. The wavy striated trochus, pearl-coloured. *Borlase Cornw. p.* 278. *tab.* 28. *fig.* 6.

T. magus tuberculated. *Penn. Br. Zool. No.* 107. *tab.* 80. *fig.* 107.

Sabot sorciere. *Argenville Conch. I. p.* 263.

Trochus pyramidalis umblicatus, anfractibus supra marginatis, infra nodulosis, albus, rubro variegatus. Tuberculatus. *Da Costa.* 25. *tab.* 3. *fig.* 1. 1.

PLATE VIII.

This Shell is found on the coasts of Sussex, Dorset, Devonshire, Cornwall, Wales, &c.—It is an elegant species, commonly white, variegated with zig-zag stripes and waves of fine red, as shewn at Fig. 1. Sometimes, however, they are of a dull yellowish tint, instead of white, with the stripes of a dark brown. The Shell is of a rich pearl colour when the outer coat is taken off.

FIG. II. III.

TROCHUS CONULUS.

CONULE.

SPECIFIC CHARACTER

AND

SYNONYMS.

Shell conic, imperforated at the base. A prominent wreath along the spires.

TROCHUS CONULUS, testa imperforata conica, lævi, anfractibus linea elevata interstinctis. *Linn. Syst. Nat. p.* 1230. *No.* 598.

Trochus pyramidalis parvus, ruberrimus, fasciis crebris exasperatus. *List. H. Conch. tab.* 616. *fig.* 2.

T. Conulus. Conule. *Penn. Br. Zool. No.* 104. *tab.* 80. *fig.* 104.

Trochus Conulus, Conule. *Da Costa, Br. Conch.* 21. *tab.* 2. *fig.* 4. 4.

Linnæus proposes this as a species, (*Conulus*); but at the same time observes, it may be a small variety of the Trochus Zizyphinus, because, like that species, it is imperforated, and has a prominent ridge on the whirls. Pennant says, it is scarcely distinct from *T. Zizyphinus.* Da Costa thinks it certainly a distinct species.

If the shell, Fig. 104. Pennant, is correct, it is of a larger growth than any of our specimens. Da Costa says, the size seldom exceeds that of a cherry kernel. Not uncommon on the shores of Sussex; and has been received from the coast of Devonshire.

Fig. II. natural fize. Fig. III. magnified.

9

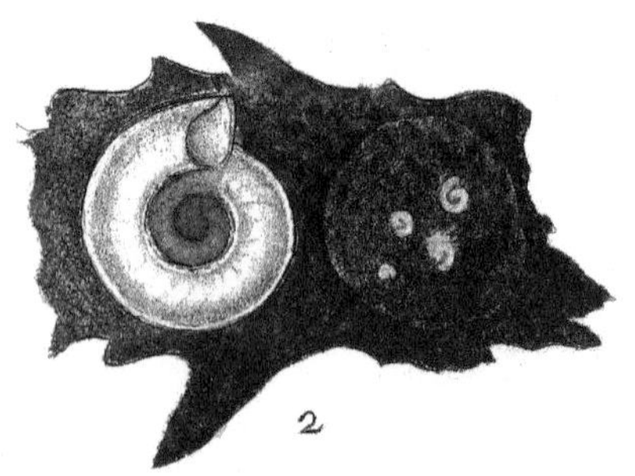

PLATE IX.

SERPULA SPIRORBIS.

GENERIC CHARACTER.

Animal a Terebella, or whimble worm. Shell tubular, adheres to other bodies, as shells, stones, &c.

SPECIFIC CHARACTER AND SYNONYMS.

Small, orbicular, spiral or wreathed like a cornu ammonis; convex above, flat beneath.

SERPULA SPIRORBIS, testa regulari spirali orbiculata: anfractibus supra introrsum subcanaliculatis sensimque minoribus.—*Gmel. Linn. Syst. Nat. Conch. p.* 3740. 5.

Vermiculus exiguus albus nautiloides, algæ fere adnascens. *List. H. Conch. tab.* 533.—*tab.* 553. *Huddesford's edition.*

Very small Worm Shells. *Dale, Harw. p.* 391. *No.* 2. *and p.* 455. *No.* 2.

Depressed orbicular Cochleæ on Algæ. *Wallis. Northumb. I. p.* 402. *No.* 41.

Serpula Spirorbis, Spiral. *Penn. Br. Zool. No.* 155. *tab.* 91. *fig.* 155.

Serpula Spirorbis, Spiral, *Vermiculaire Nautiloide.*—parva orbiculata et Spirali, ammoniæ instar convoluta. *Da Costa Br. Conch.* 12.—*tab.* 2. *fig.* 11.

PLATE IX.

This species is found in abundance on most of the British shores; it adheres to shells, stones, claws of lobsters, &c. but chiefly to the leaves of *Fucus serratus*, and other sub-marine plants. It is a strong Shell, white, and without polish; is never complicated, or laid one on another, but are dispersed singly over whatever substances they are affixed to. Petiver calls it the Wrack Spangle, because it appears like so many white spangles on the dark-coloured leaves of the Wracks.—A piece of this sub-marine plant, with the Shells adhering to it, is a very pleasing object for the opake microscope.

Fig. I. represents the natural size of the Shells. Fig. II. shews one magnified.

Obs. Dr. Lister, in his original edition, ranked this *Shell* among the *Worm-Shells* (tab. 533. fig. 5.) calling it *Nautiloides*, only from its wreathed form like to a *Nautilus*; but his re-editor, the Rev. Mr. Huddesford, has been pleased to reverse the Doctor's *arrangement*, by transposing it to the *Nautilus family*, where it now is (tab. 553), and thereby fixes an error of arrangement on *Dr. Lister*'s memory, which that excellent and accurate conchologist was not guilty of. DA COSTA, *page* 23.

10

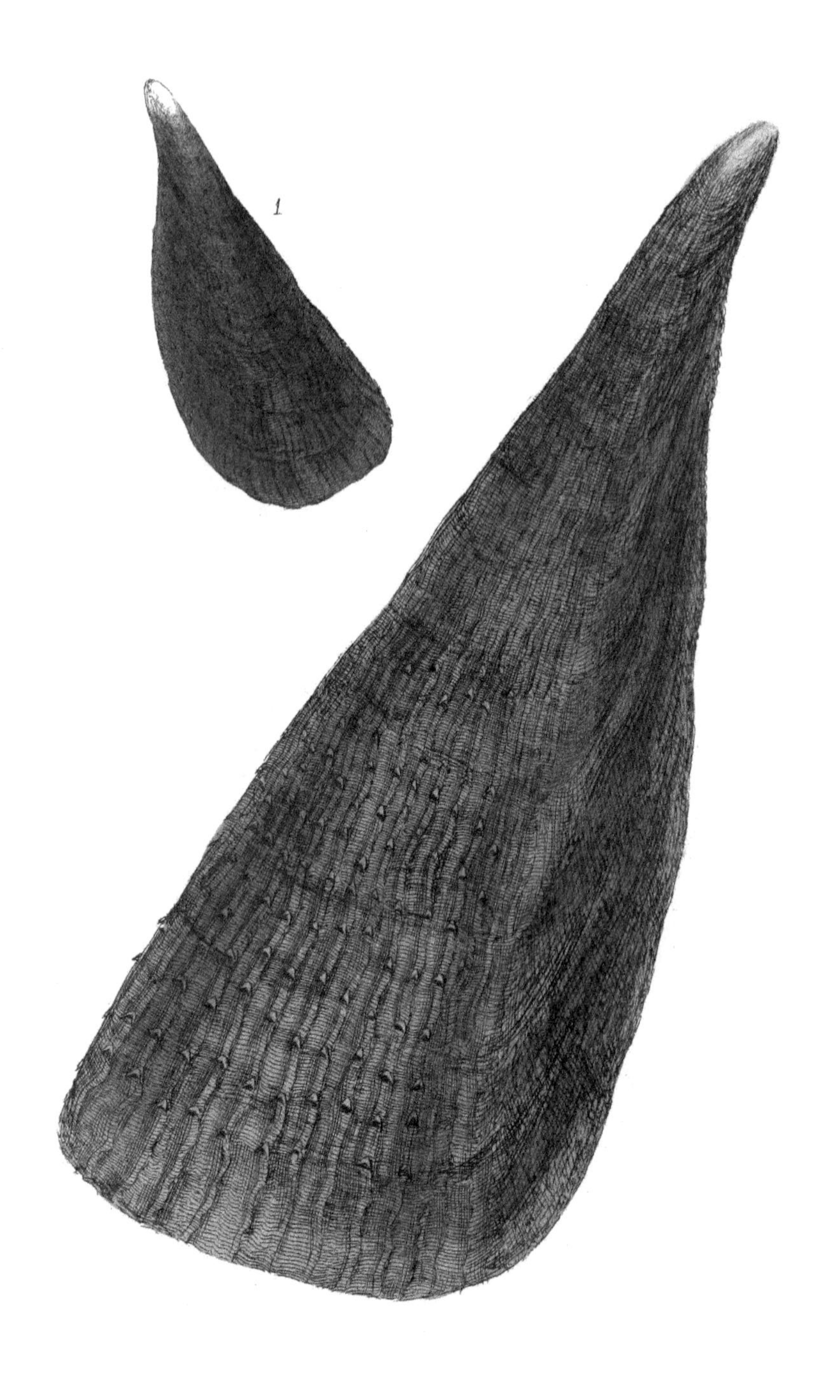

PLATE X.

PINNA MURICATA.

THORNY WING, OR SEA HAM.

GENERIC CHARACTER.

Hinge without a tooth, and placed on one side. Valves equal; open, or gape at the bottom.

SPECIFIC CHARACTER AND SYNONYMS.

Shell triangular, striated; the striæ beset with acute, ovated, and concave scales or prickles.

PINNA MURICATA: testa striata, squamis concavis ovatis acutis.
Gmel. Lin. Syst. Nat. Conch. p. 3364. *Sp.* 4.

Pinna tenuis, striata, muricata. *List. H. Conch. tab.* 370. *fig.* 210.

Pinna fragilis. Brittle. *Penn. Br. Zool. No.* 80. *tab.* 59. *fig.* 80.

Pinna tenuis costis longitudinalibus muricatis. Muricata, Thorny.
Da Costa, tab. 16. *fig.* 3. *p.* 240.

Pinna recta transversim et directe striata, et rugosa, striis in summitate aculeis exasperatis, ex fusco rubro nigricans.
Gualt. 1. *Conch. tab.* 79. *fig.* D.
Seb. Mus. 3. *t,* 92. *ser.* 1. *f.*

Concha Pinna. *Hasselq. it.* 447. *n.* 137.

Pinna lata altera. *Rumf. Mus. t.* 46. *f.* M.

PLATE X.

Dr. Rutty mentions a Pinna ten inches long and five broad, caught near the Skerries, in Ireland; and Mr. Pennant "saw specimens of vast *Pinnæ*, found among the farther *Hebrides*, in the collection of Dr. *Walker*, at *Moffat*;" but it is uncertain of what species either of these were: Mr. Pennant says, "they were very rugged on the outside, but cannot recollect whether they were of the kind found in the *Mediterranean* or West Indies*."

The only British species of Pinna we are acquainted with, is the *P. Muricata* of Linnæus, or P. Fragilis of Pennant, and that is very rare. The latter author describes it from a specimen in the PORTLAND cabinet, which had been fished up at Weymouth, in Dorsetshire. Da Costa says, he has seen a very small one (of the same species) from the coast of Wales.—Both of these are represented in the annexed plate.

This Shell is extremely thin and brittle, and gapes open at the broadest end. It is semi-pellucid, and of a horn colour; the outside marked with longitudinal ribs, roughened with rows of small prickles †. The inside is smooth, of a pale horn colour alfo, with a pearly lustre towards the top.

* This Author, however, arranges it as a new British species, without further description:—as, Pinna Ingens—Great Nacre.

† In Pennant's figure these are obsolete.

11

PLATE XI.

BUCCINUM LAPILLUS.

MASSY, OR PURPLE WHELKE.

GENERIC CHARACTER.

Aperture oval, ending in a short canal.

SPECIFIC CHARACTER.

Ovated, terminates in a sharp point, spirally ridged. Pillar lip broad.

BUCCINUM LAPILLUS: testa ovata acuta striata lævi, columella planiuscula.—*Lin. Syst. Nat. p.* 1202. *No.* 467.

Cochlea testa crassa ovata utrinque producta; spiris quinque spiraliter sulcatis; aperturæ labro undulato. *Faun. Suec. p.* 378. *No.* 2167.

Buccinum minus, albidum, asperum, intra quinas spiras finitum. *List. H. An. Ang. p.* 158. *tit.* 5. *tab.* 3. *fig.* 3.

Buccinum brevi rostrum supra modum crassum, ventricosius, labro denticulato: Purpura Anglicana. *List. H. Conch. tab.* 965. *fig.* 18.—ET B. brevi rostrum, album denticulo unico ad imam columellam. Purpura Anglicana. *Fig.* 19.

Purple marking Whelke. *Borlase Corn. p.* 277. *tab.* 28. *fig.* 11.

English purple. *Smith Cork. p.* 318.

Horse wrinkles. *Smith Waterford. p.* 272.

Small purple Whelke. *Wallis Northumb. p.* 401.

Buccinum lapillus, Massy. *Penn. Br. Zool.* 4. *No* 89. *tab.* 72. *fig.* 89.

PLATE XI.

Buccinum canaliculatum minus, crassum varicolor, striatum, seu Purpura Anglicana. Purpuro-buccinum. *Da Costa Br. Conch. tab.* 7 *fig.* 1. 2. 3. 4. 9. 12.

This is a strong, thick shell, generally about one inch and a half in length, of a full pyramidal shape, with a point acute; it has five spires, furrowed: the ridges of the lower wreath notched, or scaled, and very rough. Within the mouth it has five long parallel teeth.

The colours are various, often of a simple and uniform yellowish brown, sandy, or clay colour; sometimes quite white, or white tinged with violet, and fasciated with yellow or brown; the latter are the most elegant varieties of B. Lapillus.—These shells are found in great abundance near low water-mark, on many of the shores of Great-Britain. It is one of the species that yields the purple dye analogous to the *purpura* of the ancients; and though the value of its dye has been long superseded by the cochineal insect, the shells that produced it are objects of curiosity. The Tyrian purple was the most admired, and is known to have been extracted from a species of the *Murex*; but other purples of inferior lustre are also mentioned by the ancients. Da Costa imagines that the liquor of this Whelke (Buccinum Lapillus) was a valuable purple to the ancient English, and quotes the authority of *Bede*, who lived about the seventh century, for this opinion. "There are," says *Bede*, "snails in very great abundance, from which a scarlet or crimson dye is made, whose elegant redness never fades, either by the heat of the sun, or the injuries of rain, but the older it is, the more elegant*."

* Sunt cochleæ, satis superque abundantes, quibus tinctura coccinei coloris conficitur. Cujus rubor pulcherrimus nullo unquam solis ardore, nulla valet pluviarum injuria pallescere; sed quo vetustior, eo solet esse venustior.—*Bede, Hist. Eccles.* (edit. opt.) l. i. c. i. p 277.

PLATE XI.

In 1684, Mr. Cole, of Bristol, described the process of extracting the *purple* of this shell, in the Philosophical Transactions. His account is as follows:

" The Shells being harder than most of other kinds, are to be broken with a smart stroke with a hammer, on a plate of iron, or firm piece of timber (with their mouths downwards) so as not to crush the body of the fish within; the broken pieces being picked off, there will appear a white vein, lying transversely in a little furrow, or cleft, next to the head of the fish, which must be digged out with the stiff point of a horse-hair pencil, being made short and tapering. The letters, figures, or what else shall be made on the linnen (and perhaps silk too) will presently appear of a pleasant light green colour, and if placed in the sun, will change into the following colours, i. e. if in winter, about noon; if in summer, an hour or two after sun-rising, and so much before setting; for, in the heat of the day in summer, the colours will come on so fast, that the succession of each colour will be scarcely distinguished. Next to the first light green, it will appear of a deep green, and in a few minutes change into a sea-green; after which, in a few minutes more, it will alter into a watchet-blue; from that, in a little time more, it will be of a purplish-red; after which, lying an hour or two, (supposing the sun still shining) it will be of a very deep purple-red, beyond which the sun can do no more.

" But then the last and most beautiful colour, after washing in scalding water and soap, will (the matter being again put into the sun or wind to dry) be of a fair bright crimson, or near to the prince's colour, which, afterwards, notwithstanding there is no use of any stiptick to bind the colour, will continue the same, if well ordered, as I have found in handkerchiefs that have been washed more than forty times; only it will be somewhat allayed from what it was after the first washing. While the cloth so writ upon lies in the sun, it will yield a very strong and fœtid smell, as if garlick and assafœtida were mixed together."

12

1

2

PLATE XII.

OSTREA SUBRUFUS.

GENERIC CHARACTER.

Animal a Tethys. Shell bivalve unequal. The hinge without a tooth, having a small oval cavity.

SPECIFIC CHARACTER AND SYNONYMS.

Shell thin. Twenty longitudinal rays, finely striated; ears unequal; colours various; generally red.

PECTEN TENUIS, subrufus, maculosus, circiter viginti striis majoribus, at lævibus, donatus. *List. H. An. Angl. p.* 85. *tab.* 5. *fig.* 30.

PECTEN SUBRUFUS. *Penn. Br. Zool. No.* 63. *tab.* 60. *fig.* 63.

PECTEN PICTUS: mediocris, fere æquivalvis, tenuis, variis coloribus perbelle variegatus. *Da Costa. Br. Conch. p.* 144. *ſp.* 3.

Pectunculus pennatus striis dense notatus, luteo purpurascens. Pecten altis striis albo purpureis transverse variegatis insignis; & Pectunculus purpurascens vittis albis circularibus variegatus. *Borlase Cornw. p.* 277. *tab.* 28. *fig.* 18, 21 *and* 22.

PLATE XII.

This elegent species is found on several of the shores of Great Britain and Ireland, particularly those of Cornwall, Dorset, and Northumberland. It is generally about two inches and an half in length. Shell thin and rather convex. The inside is smooth and glossy, and commonly white, though sometimes of a brownish colour. The colours of the outside very various and beautiful. Da Costa enumerates the chief varieties, as, 1. *almost white*, and *white* charged with *brown*, *red*, or *purple*; 2. *uniform* bright *yellow*, and pale *yellow*, with *white*; 3. *uniform brown*, and *brown*, *red*, or *purplish* grounds with *white*, &c. all these colours are elegantly blended and variegated, sometimes marbled or mottled or disposed in *zones*, *girdles*, broad longitudinal rays, &c.

Fig. 1. represents a fine coloured specimen of the variegated red and white kind. Fig. 2. The uniform deep orange, which we apprehend is less common.

13

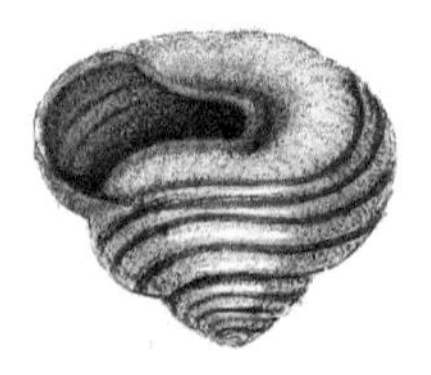

PLATE XIII.

HELIX NEMORALIS.

GIRDLED SNAIL.

GENERIC CHARACTER.

Aperture or mouth contracted and lunated.

SPECIFIC CHARACTER

AND

SYNONYMS.

Imperforated, subrotund, thin, pellucid. Mouth semi-lunar; generally girdled with streaks: and of various colours.

HELIX NEMORALIS: testa imperforata subrotunda lævi diaphana fasciata, apertura subrotundo-lunata. *Linn. Faun. Suec.* 2186.—*Gmel. Linn. Syst. Nat. Conch. p.* 3647. 108.

Cochlea citrina aut leucophæa, non raro unicolor, interdum tamen unica, interdum etiam duobus, aut tribus, aut quatuor plerumque vero quinis fasciis pullis distincta. *List. H. An. Angl. p.* 116. *tit.* 3. *tab.* 2. *fig.* 3.

Cochlea imperforata, interdum unicolor, interdum variis fasciis depicta. FASCIATA girdled. *Da Costa, Br. Conch. p.* 76. *sp.* 41.

Helix Nemoralis, variegated. *Penn. Br. Zool. No.* 131.

F 2

PLATE XIII.

Prof. Gmelin, in the last edition of the Systema Naturæ, enumerates no less than thirty-one varieties of this beautiful land Shell. Da Costa describes six principal varieties in his British Conchology *. Some of the kinds are rare, others extremely common, living in trees, hedges and gardens. It is a widely diffused species being found in every part of Europe as well as Great Britain.

* 1. *Uniform*, of a *pale citron* colour, or *yellow* of *different shades*: the *mouth* finely bordered within and without, with a dark brown, and with a brownish shade or cloud on so much of the body wreath as lies within the mouth, or from the outer lip quite across to the edge of the pillar. Pretty *frequent*.

2. *Uniform*, of a *flesh colour* of *different shades*, with the mouth in like manner bordered with dark brown; and the body wreath also shaded exactly the same as the last. Not very *frequent*.

3. *Uniform*, of *different degrees* of *brown*, with the same circumstances. Common.

4. The *ground yellow* or *greenish yellow* of *different shades*, with a regular *single* ſpiral *girdle*, or according to the turn of the wreaths, in the very *middle* of each wreath, with the brown border round the *mouth*, and the shade or cloud on the *body*. Pretty frequent.

5. *The ground fleſh colour* of *different shades*, variegated in like manner with a *single girdle*, the *border* round the *mouth*, and on the *body*. *Not very frequent*.

6. *Many* dark-brown spiral *girdles* on the *yellow*, *fleſh*, or *brownish grounds*, sometimes to *five girdles* at least on the body wreath; sometimes only *four*. Theſe *girdles* are of different *breadths*, some being very *narrow*, like streaks, others broader, like belts; and others *so extremely broad* as to cover the parts, and make the *ground colour* only appear in *girdles*. They are also not equidistant or regularly set; but *the very broad girdles* lie most generally on the upper part of the shells. These *girdled sorts* are the *most frequent* or *common*. Da Costa *Br. Conch. p.* 78.

14

PLATE XIV.

PATELLA VULGATA.

COMMON LIMPET.

GENERIC CHARACTER.

Animal Limax. Shell univalve, subconic, without spires.

SPECIFIC CHARACTER

AND

SYNONYMS.

Oblong ovoid with about fourteen obsolete angles, margins deep or dilated.

PATELLA VULGATA: testa subangulata: angulis quatuor decim obsoletis margine dilato acuto.—*Gmel. Linn. Syst. Nat. Conch. p.* 3697.—*Sp.* 23 β *Schroet. n. Litterat* 3. *p.* 62. *n.* 117 γ *Knorr. Vergn* 6. *t.* 27. *f.* 8.

Patella integra ex livido cinerea, striata. DA COSTA. *Br. Conch. p.* 3. *pl.* 1. *fig.* 1, 2, 8.

Patella ex livido cinerea striata. *List. Hist. Anim. Angl. p.* 195. *tit.* 40. *tab.* 5. *fig.* 40.

Patella Vulgata, Common. *Penn. Br. Zool.* 4. *No.* 145. *tab.* 89. *fig.* 145.

Patella integra. *Klein, Ostracol. p.* 115. §. 283. *No.* 10.

Lepas *Argenville, p.* 21.

F 3

PLATE XIV.

The Limpet is common on all the European ſhores. The outside is generally encrusted with filth, balani, &c. beneath which, it has an epidermis of a blackish colour. The shells vary exceedingly in colours, not only in the different stages of growth, but also in the adult state. When young, the colours are remarkably vivid and elegantly disposed; the shell flat and the margins deeply crenated: those of full growth are on the contrary very conic and the colours less brilliant. The margins irregular and the ridges more obsolete. Some authors have considered several varieties as distinct species. Da Costa among others, deems the *Patella depressa* of Pennant, no other than a young variety of the common kind.

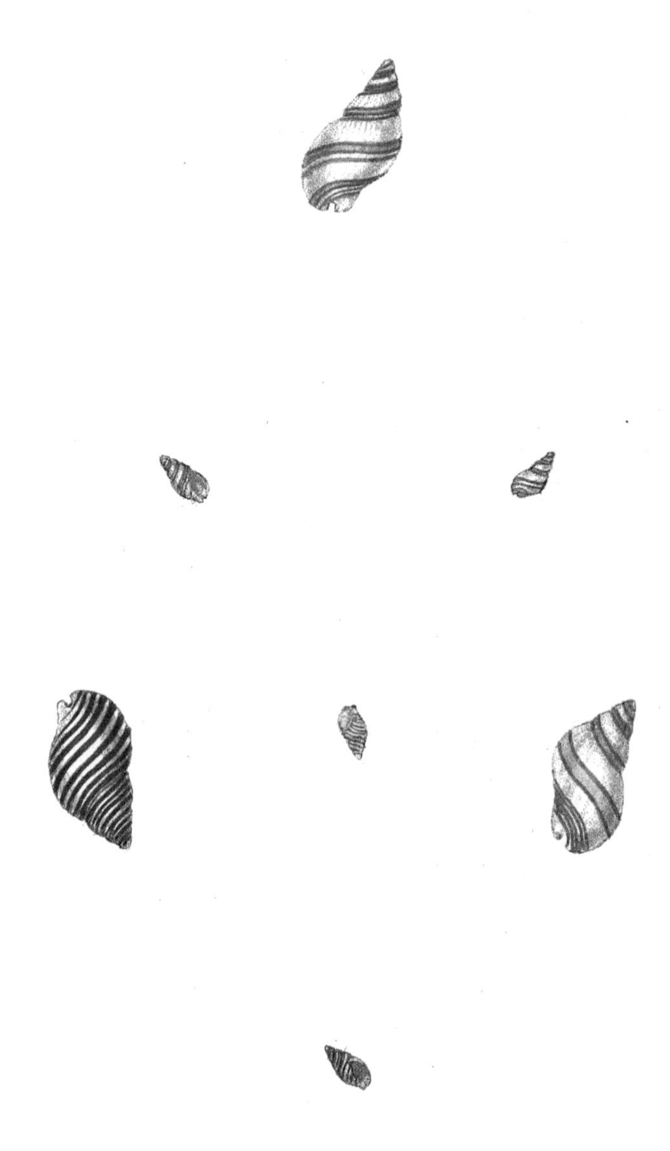

PLATE XV.

BUCCINUM LINEATUM,

LINEATED.

GENERIC CHARACTER.

Whelkes whose mouths are cut short at top, for the gutter or beak does not ascend, but bends and falls on the back, oblique or awry, exactly like the mouth of a soal or flat fifh. *Da Costa.*

SPECIFIC CHARACTER.

Small, pyramidal, or sharp pointed at bottom. Dark brown, lineated fpirally with white.

BUCCINUM LINEATUM: recurvirostrum minimum pullum, lineis albidis spiraliter distinctum. *Da Costa, Br. Conch. p.* 130. *sp.* 77.

This species is found in great abundance on the coast of Cornwall. The annexed plate exhibits several magnified figures of the most elegant varieties, together with the natural size.

2
1
2

PLATE XVI.

FIG. I.

NERITA PALLIDULUS.

PALE NERIT.

GENERIC CHARACTER.

Globose. Aperture semiorbicular.

SPECIFIC CHARACTER

Semitransparent. Wreaths rather prominent. Mouth semilunar, very patulous. Umbilicus large.

Nerita Corneus, spira paululum exserta. Pallidulus *Da Costa. p.* 51. *Sp.* 29.

Da Costa says " This species is rare, for I have only received some few shells from the coasts of Kent and Dorset." He also considers it an undescribed shell.

PLATE XVI.

FIG. II. II.

NERITA FLUVIATILIS.

RIVER NERIT.

SPECIFIC CHARACTER,

AND

SYNONYMS.

Small, spotted, streaked, and reticulated.

NERITA FLUVIATILIS; N. testa rugosa labiis edentulis.—*Linn. Syst. Nat. p.* 125. 3. *No.* 723.

Nerita parvus fluviatilis, elegantur maculatus, fasciatus, aut reticuculatus. Flaviatilis. *Da Costa Br. Conch. p.* 48. *Sp.* 27.

Nerita fluviatilis, é cœruleo virescens, maculatus, operculo subrufo lunato et aculeato datus. *List. H. An. Angl. p.* 136. *tit.* 20. *tab.* 2. *fig.* 20.

Nerita fluv. exiguus, recticulate variegatus. Small netted Thames nerit. *Muſ. Petiv. p.* 67. *No.* 718.

Nerita fluviatilis, River. *Penn. Br. Zool. No.* 142. *tab.* 87. *fig.* 142.

This species is very frequent in rivers. It is small; of an ovoid shape, and very elegantly variegated with black, white, red, green, &c.—The star denotes the natural size of the shell.

17

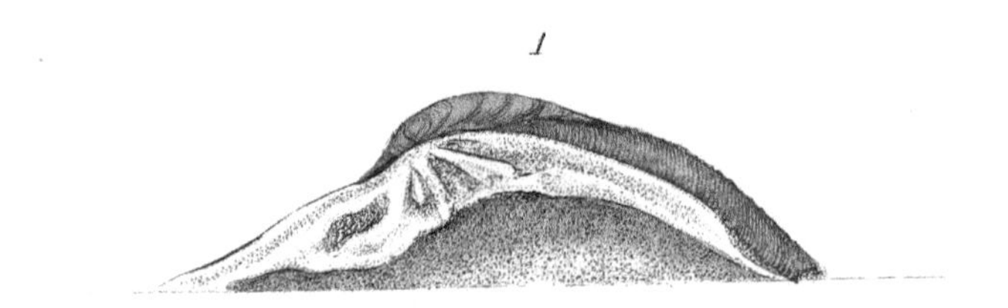

PLATE XVII.

VENUS CHIONE.

GENERIC CHARACTER.

Bivalve. Hinge furnished with three teeth; two near each other, the third divergent from the beaks.

SPECIFIC CHARACTER AND SYNONYMS.

Shell smooth with fine transverse wrinkles, a strong cartilage on one slope, and a long pointed oval depression on the other.

VENUS CHIONE: testa transverse subrugosa lævi, cardinis dente posteriori lanceolato. *Gmel. Linn. Syst. Nat. Conch. p.* 3272. *sp.* 16.

P. GLABER, SMOOTH Pectunculus major crassus, politus, castaneus, lucide radiatus. *Da Costa. Br. Conch. p.* 184. *sp.* 22.

Pectunculus maximus crassus, lævis fere radiatus. *Mus. Petiv. p.* 86. *No.* 833.—*Curvirostrum. Leigh. Lancashire. tab.* 3. *fig.* 5.

Venus Chione, β *Rumf. Mus. t.* 42. *f.* G.

Venus Chinone, γ *Chemn. Conch.* 6. *t.* 33. *f.* 334.

" This species," says Da Costa, " is rare in England. I found it at Mount's Bay in Cornwall, where the fishermen told me they call it *Queen Fish;* it is also found near *Fowey* and other shores of that county. I have seen some from *Weymouth*, and Mr. Petiver received it from the iſland of Purbeck, in Dorsetshire. Dr. Leigh mentions that it is got on the coasts of Cheshire."

Pennant has not noticed this Shell. Linnæus described it as an Asiatic species in the Systema Naturæ, but adds it is perhaps an European species also. In the last edition by Gmelin, it stands expressly as a British Shell. *Habitat in Mari Britannico, &c.*

This Shell is thick, strong and heavy: the outside smooth and glossy, with numerous concentric transverse wrinkles, and several faint rays in a longitudinal direction. The margins are plain. The inside milk white and glossy.

18

PLATE XVIII.

TURBO FASCIATUS.

FASCIATED.

GENERIC CHARACTER.

Animal Limax. Univalve, spiral, or of a taper form. Aperture somewhat compressed, orbicular, entire.

SPECIFIC CHARACTER AND SYNONYMS.

Six spires. White marbled or fasciated with black.

TURBO FASCIATUS. Fasciated. *Penn. Br. Zool. No.* 119. *tab.* 82. *fig.* 119.

Buccinum exiguum fasciatum et radiatum. *List. H. Conch. tab.* 19. *fig.* 4.

This is one of the six species Da Costa marks with a roman letter, because he could not procure the originals to figure and describe in the British Conchology. It is figured in the British Zoology of Pennant, who says it is very frequent in *Anglesea*, in sandy soils near the coast.

19

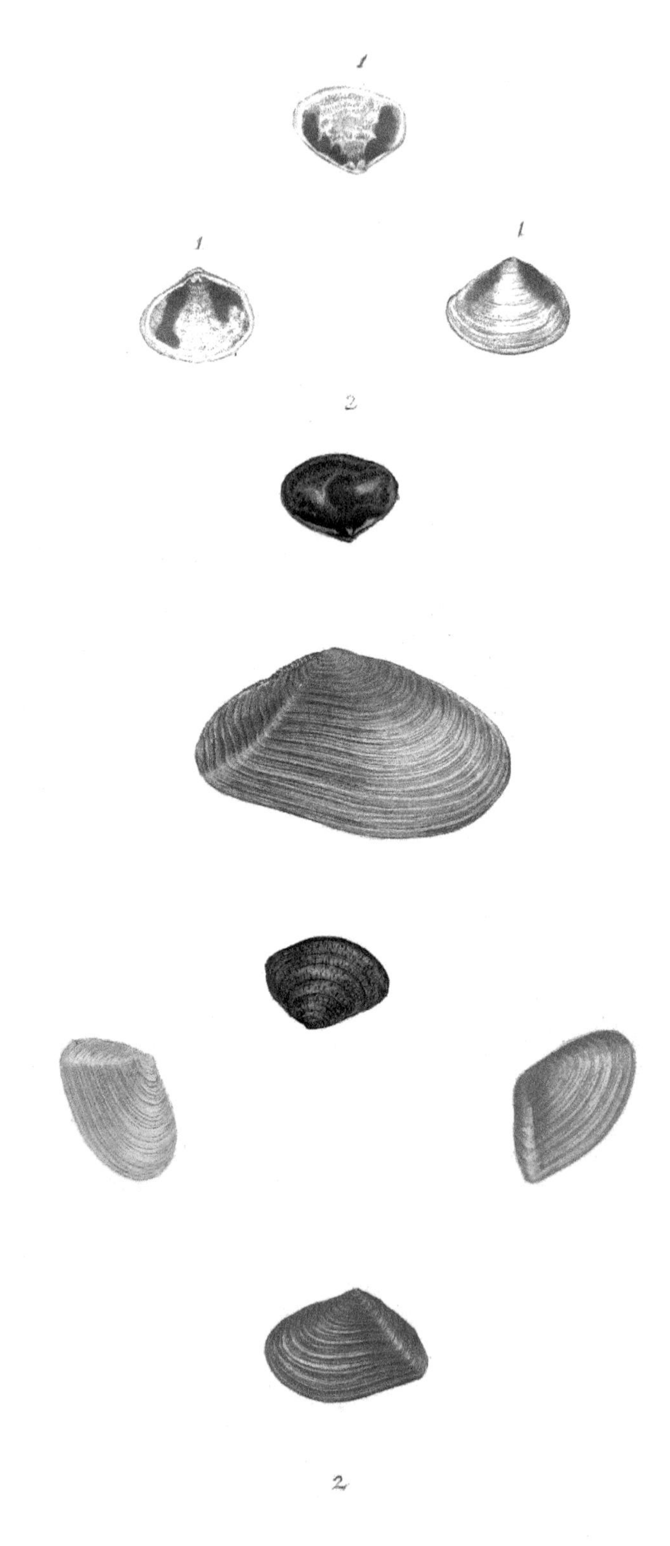

PLATE XIX.

FIG. I. I.

TELLINA BIMACULATA.

DOUBLE SPOT TELLEN.

GENERIC CHARACTER.

The hinge usually furniſhed with three teeth. Shell generally sloping on one side.

SPECIFIC CHARACTER.
AND
SYNONYMS.

Subrotundand somewhat triangular, smooth and whitish. Two oblong sanguineous red spots on the inside.

TELLINA BIMACULATA: testa triangulo-subrotunda latiore lævi albida: intus maculis duabus sanguineis oblongis. *Linn. F. Suec.* 11. *No.* 2135.—*S. N. p.* 1120.

T. minima lævis alba, intus maculis duabus sanguineis oblongis notata. Binaculata. *Da Costa. Br. Conch. p.* 213. 45.

This singular species is found on the shores of Lancashire and Hampshire. 1

PLATE XIX.

FIG. II. II.

TELLINA TENUIS.

THIN TELLEN:

SPECIFIC CHARACTER
AND
SYNONYMS.

Thin. Subrotund, glossy ;—colour sometimes red

TELLINA PLANATA: Concha testa subrotunda glabra incarnatà. *Linn. F. Suec.* 1. *p.* 381. *No.* 1335.

Tellina testa-ovata compressa, transversim substriata lævi: marginibus acutis, &c. *S. N. p.* 1117. *No.* 52.

Tellina valde tenuis, parva, subrotunda, plerumque rubra. Tenuis. Thin. *Da Costa. Br. Conch. p.* 210. *Sp.* 43.

Tellina parva, intus rubra, ad alterum latus sinuosa. *List. Conch. tab.* 405. *fig.* 250.

Tellina lævis intus et extra rubra, ad latus sinuosa. *Ib. fig.* 251.

Tellina planata. *Plain. Penn. Br. Zool. No.* 29. *tab.* 48. *fig.* 29.

Found on many of our shores, as Kent, Essex, Cornwall, &c. Some of the varieties are extremely delicate, and prettily streaked with pale red and white: many are entirely white, or white tinged with yellow. Some are orange colour; but the rarest kind is deep violet or purple.

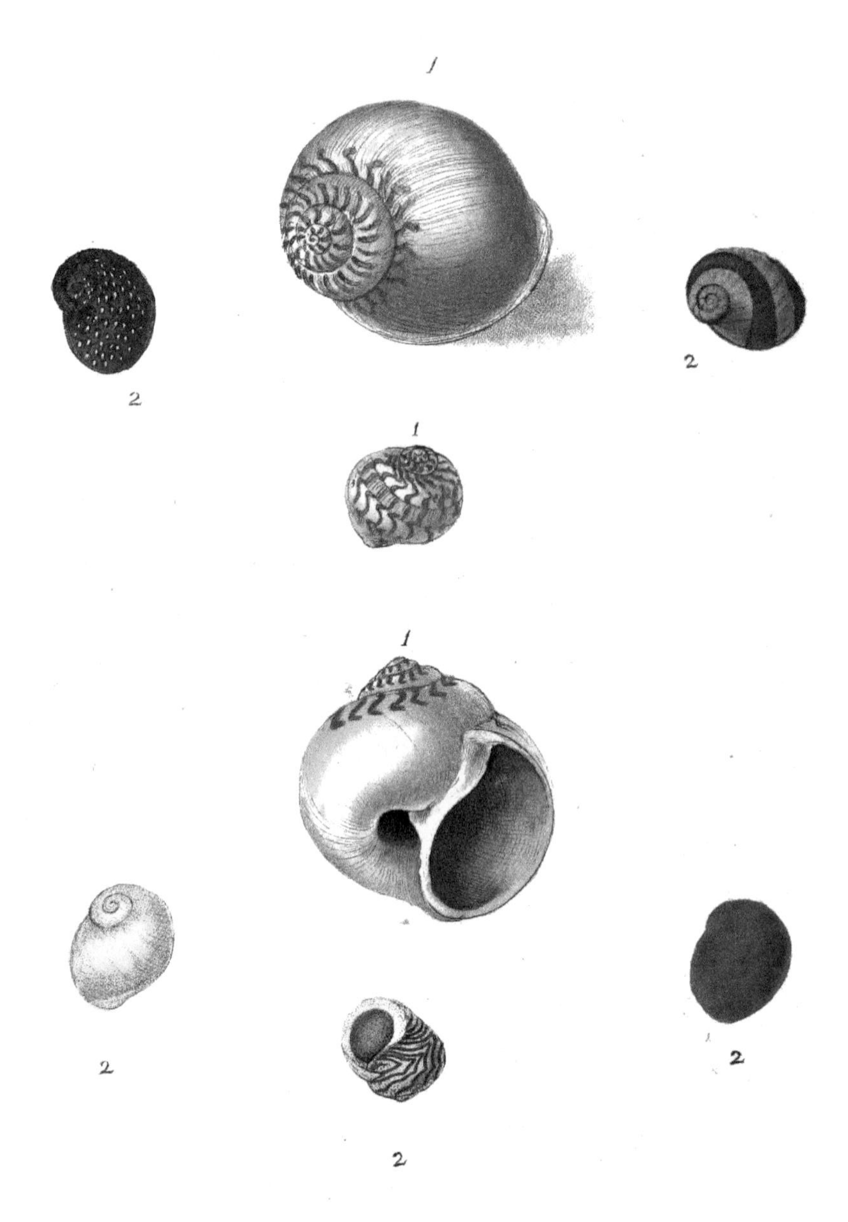
1
2
2
1
1
2
2
2
2

PLATE XX.

FIG. I. I.

NERITA GLAUCINA.

CHAIN NERIT.

GENERIC CHARACTER.

Globose. Aperture semiorbicular.

SPECIFIC CHARACTER

AND

SYNONYMS.

Umbilicated, glossy. Spires swelled, obtuse. Umbilicus large and deep with the inner lip greatly spread on the body wreath. A chain of short brown marks along the spires.

NERITA GLAUCINA: testa umbilicata lævi, spira obtusiuscula, umbilico semiclauso: labio gibbo dicolore. *Linn. Syst. Nat. p.* 1251. *No.* 716.

Nerita Glaucina. Livid. *Penn. Br. Zool. No.* 141. *tab.* 87. *fig.* 141.

Cochlea Catena. Chain Nerit. C. Umbilicata albo rufescens fasciis maculatis, maxime ad imos orbes distincta. *Da Costa, Br. Conch. p.* 83. *sp.* 45.

This Shell is not uncommon on the shores of the Essex and Kentish coasts; on the sandy shores of Lincolnshire, Dorsetshire, Devonshire,

Cornwall, &c. The same species is found in the Mediterranean and the West Indies. The colours are very fine, particularly in the young Shells.

FIG. II. II.

NERITA LITTORALIS.

SPECIFIC CHARACTER AND SYNONYMS.

Shell thick, smooth vertex flat. Inner lip spread obliquely.

NERITA LITTORALIS : T. teſta lævi, vertice carioso, labiis edentulis.
Lin. Syst. Nat. p. 1253. *No.* 724.
Nerita vulgaris, unicolor, flavus aurantiacus, vel fuscus, aut fasciatus, aut reticulatim variegatus. Littoralis.
Da Costa, Br. Conch. p. 50. *sp.* 28.
List. H. Conch. tab. 697, *fig.* 39.
Nerita Littoralis. Strand. *Pen. Br. Zool. No.* 143. *tab.* 87. *fig.* 143.

This Shell is very common on all the British coasts, particularly the fine yellow kinds. Those with broad bands or girdles, and also such as are reticulated with dark greenish colour on a light ground, are rare varieties of this species.

3

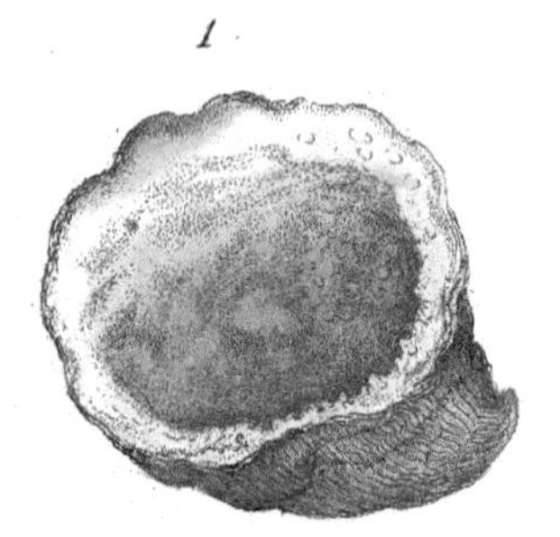
1

1

2

3

PLATE XXI.

PATELLA HUNGARICA.

LARGE FOOL'S CAP.

GENERIC CHARACTER.

Animal Limax. Shell univalve, subconic, without spires.

SPECIFIC CHARACTER

AND

SYNONYMS.

Shell entire, conic, acuminated, striated, with the vertex turning down, or hanging over one side.

PATELLA UNGARICA: testa integra conico acuminata striata vertice hamoso revoluto. *Linn. Syst. Nat. p.* 1259. *No.* 761.

PATELLA HUNGARICA. Bonnet. *Penn. Br. Zool. No.* 147. *tab.* 90. *fig.* 147.

Patella integra, albescens, striata, vertice spirali, intus rosacea. *Da Costa, Br. Conch. p.* 12. *sp.* 6.

Lepas Bonnet de Dragon. *D'Avila, Cab. I. p.* 86. 87. *No.* 32. 34.

"This species is only found on the *Cornish* coast, and even is very scarce there, being most generally dredged some miles from the shore; for the Shell is so thin, that it will hardly bear rolling from its native spot to the beach. It is generally found affixed to a species of *escallops*, called *frills*, in Cornwall." *Da Costa.*

PLATE XXI.

FIG. II. II.

PATELLA PARVA.

SMALL LIMPET.

SPECIFIC CHARACTER.

Shell small, entire, without gloss, whitish, faintly rayed with red.

PATELLA PARVA: integra, parva, sublævis, albescens radiis rubentibus. *Da Costa Br. Conch. p. 7. sp.* 3.

Da Costa considers this as a nondescript species; he received several specimens of it from the coasts of Dorsetshire, but never from any other of the British shores, and therefore proposes it as a scarce Shell.

It is rather larger than a pea, thin, and semipellucid; of a depressed conic shape, and the vertex inclining very much to one side. The inside is whitish, outside the same, with a few longitudinal rays of pale red, or purplish brown.

PLATE XXI.

FIG. III. III.

PATELLA RETICULATA.

RETICULATED MASK LIMPET.

SPECIFIC CHARACTER
AND
SYNONYMS.

Small, ash colour, reticulated. Vertex perforated.

P. LARVA RETICULATA. Patella parva cinerea, vertice perforata. *Da Costa Br. Conch. p.* 14. *sp.* 7.

PATELLA GRÆCA. STRIATED. *Penn. Br. Zool. No.* 153. *tab.* 89. *fig.* 153.

PATELLA CLATHRATA. *Klein. Ostrac. p.* 116.—284. *No.* 2. *List. H. Conch. tab.* 527. *fig.* 2.?

Pennant says, this species inhabits the west of England. The specimens in Da Costa's collection were also fished up near Weymouth, in Dorsetshire. This is a rare Shell, and is not known to inhabit any other of the British coasts.

This Shell is about three quarters of an inch in length, half an inch in breadth, and one quarter of an inch in heighth. The outside is deeply reticulated, or wrought with prominent longitudinal and transverse ridges. The vertex inclines to one end, and is perforated; its aperture is of an oblong form, and about one tenth of an inch in length.

2
1
1
2
2

PLATE XXII.

FIG. I. I.

TURBO CINCTUS.

GIRDLED WREATH SHELL.

GENERIC CHARACTER.

Animal Limax. Univalve, spiral, or of a taper form. Aperture somewhat compressed, orbicular, entire.

SPECIFIC CHARACTER
AND
SYNONYMS.

Whitish, variegated with brown. Spires swelled ana ridged: two particularly large, broad, roundish ridges, in the middle of each spire.

Turbo Cinctus: strombiformis medius albus pullo variegatus, anfractibus porcis tumidis latis & spirabilibus cinctus. *Da Costa Brit. Conch. p.* 114. *sp.* 66. *Tab.* 7. *fig.* 8.

Turbo Exoletus. *Linn. Syst. Nat.* ?

This is a very rare Shell. Da Costa says he has received it only from the coasts of Lincolnshire and Lancashire.

H 2

PLATE XXII.

FIG. II. II. II.

TURBO TEREBRA.

AUGER SHELL.

SPECIFIC CHARACTER
AND
SYNONYMS.

Shell slender. Spires twelve, striated spirally. Six of the *striæ* rather prominent.

TURBO TEREBRA: testa turrita anfractibus carinis sex acutis. *Gmel. Linn. Syst. Nat. Conch. p.* 3608. *sp.* 81.

Cochlea testa longa subulata, spiris, duodecim striatis. *Linn. Faun. Suec.* 1. *p.* 378. *No.* 1322. 2. *No.* 2171.

Buccinum tenue, dense striatum, duodecim minimum spiris donatum. *List. H. An Angl. p.* 161. *tit. tab.* 3. *fig.* 8.

Strombiformis medius albus rufo variegatus, anfractibus striatis. Terebra. *Da Costa Brit. Conch. p.* 112. *sp.* 65.

Turbo Terebra. Auger. *Penn. Br. Zool. No.* 113. *tab.* 81. *fig.* 113.

The colours in this species vary exceedingly; the ground colour is generally white, or cream colour, with the streaks, dots, and markings of brown, pale red, or orange. The length is from one inch and an half to two inches or more.

PLATE XXII.

It is not uncommon on many of the British coasts *. Adanson has a variety of it (β) from Senegal ; and other authors mention the same species as a native of the East Indian and African seas.

* This species is not uncommon on many of our coasts, and in great plenty on some, as at the *Scilly Islands* ; at *Liverpool*, where they are called Cockspurs ; at *Scarborough*, after winter storms, according to Lister ; at *Exmouth*, and other places on the western shores; and I have received very fine and perfect ones from the coasts of Wales, as *Flintshire*, *Pwlbely* in *Carnarvonshire*, and *Barmouth* in *Merionethshire*. It is also a Shell of the Orkneys. *Da Costa*.

23

PLATE XXIII.

MYTILUS MODIOLUS.

GREAT MUSCLE.

GENERIC CHARACTER.

The hinge toothless, and consists of a longitudinal furrow.

SPECIFIC CHARACTER.

AND

SYNONYMS.

Shell large, blackish: one side angulated near the middle, the other straight; but gibbous towards the beaks, and blunted or obtuse at the upper end.

MYTILUS MODIOLUS: testa lævi, margine anteriore carinato, natibus gibbis, cardine sublaterali.—*Gmel. Linn. Syst. Nat. Conch. p.* 3354. *Sp.* 14.

Mytilus magnus nigrescens. Modiolus. *Da Costa. Br. Conch. p.* 219. *sp.* 49. *tab.* 15. *fig.* 5.

Musculus papuanus authorum. *Rumph. Mus. tab.* 46. *fig. B.*

M. Modiolus, Great Mufcle. *Penn. Br. Zool. p.* 113. 77. *tab.* 46. *fig.* 77.

Musculus papaunus. *Adans. Seneg.* 1. *t.* 22. *f. C.*

List. H. Conch. tab. 359. *fig.* 198.

Gualt. test. t. 91. *H. L.*

Rumph. Mus. t. 46. *f. B. C? D?*

H 4

PLATE XXIIII.

Mytilus Modiolus is the largest species of this genus that inhabits the British shores; being from six to seven inches in length, and three in breadth. It is a strong and heavy shell; the outside is of a blackish colour inclining to purple. It is covered with a thin filmy brown epidermis, and often with balani and other remains of crustaceous animals. Within, it is smooth and pearly, and sometimes richly coloured with a variety of vivid hues, in which red, purple and green chiefly predominate. These shells lie only in deep waters, and are never cast upon shore; but sometimes they seize the bait of the ground lines, and are hauled up by the fishermen.

Da Costa received the M. Modiolus, of a ſmall size, from the Margate flats in Kent; from Cornwall and other English shores. The specimen figured in the annexed Plate is from Scarborough in Yorkshire; thoſe found on the coast of Wales and Scotland, and particularly the Orkneys, are not inferior in point of size to those from Scarborough.

24

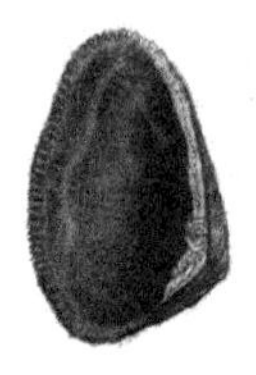

PLATE XXIV.

DONAX CRENULATA.

PURPLE or TRUNCATED PURR.

GENERIC CHARACTER.

Bivalve. One side very obtuse, margin crenated. Hinge various, generally of two teeth.

SPECIFIC CHARACTER AND SYNONYMS.

One side very blunt or truncated: thickly striated longitudinally. Margin serrated.

DONAX RUGOSA: testa antice rugosa gibba, marginibus crenatis. *Linn. Syst. Nat. p.* 1127. *No.* 104. *Mus. reg. p.* 494. *No.* 50.

DONAX DENTICULATA: testa anterius obtusissima: labiis transverse rugosis, margine denticulato, nymphis dentiformibus. *Gmel. Linn. Syst. Nat. Conch. p.* 3263. *sp.* 6.

Cuneus ex albo & violaceo radiatus, intus vero violaceus, latere altero gibbo & truncato. TRUNCATUS. Truncated Purr. *Da Costa, Br. Conch. p.* 205. *sp.* 40.

Tellina intus ex viola purpurascens, in ambitu serrata. *List. Hist. An. Angl. p.* 190. *tit.* 35. *tab.* 5. *fig.* 35.

PLATE XXIV.

Tellina crassa, admodum leviter striata, intus violacea. ***List. H.*** *Conch. tab.* 375. *fig.* 216.—376.—218. 219.

DONAX DENTICULATA. Purple. *Penn. Br. Zool. No.* 46.

This is a very elegant and remarkable species: the annexed Plate exhibits five of its most singular varieties. The young shells are sometimes quite white, or white faintly marked with brown, red or violet; the old shells are of a deep violet without, and variously marked with the same on the outside. Very common on the western coasts of England, and also on those of Ireland and Scotland.

25

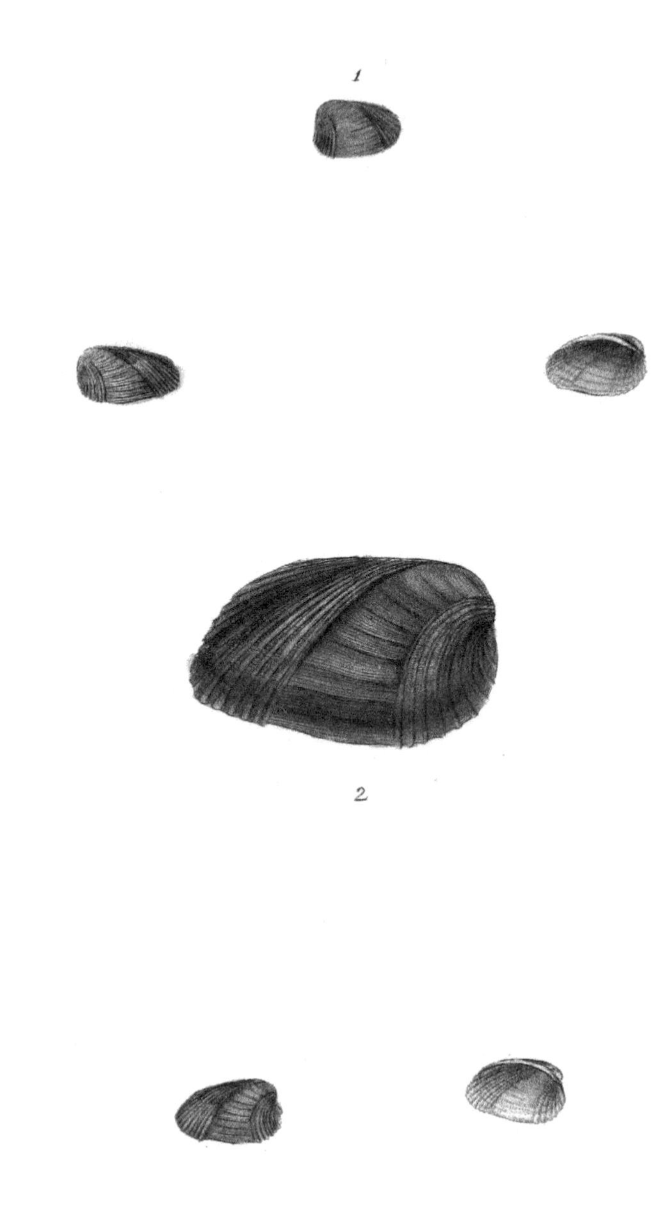

PLATE XXV.

FIG. I. I.

MYTILUS DISCORS.

DIVIDED MUSCLE.

GENERIC CHARACTER.

The hinge toothlefs, and consits of a longitudinal furrow.

SPECIFIC CHARACTER.
AND
SYNONYMS.

Shell oval, somewhat diaphanous and horny. Outside divided into three compartments ; the two exterior striated longitudinally ; the middle transversely, with extremely fine striæ.

MYTILLUS DISCORS : testa ovali cornea subdiaphana, antice longitudinaliter posterius transversaliter striata.— *Gmel. Linn. Syst. Nat. Conch. p.* 3356. *sp.* 21.

Mytillus minor tenuis areis tribus distinctus. Discors. *Da Costa, Br. Conch. p.* 221. *sp.* 51.

The discovery of this rare species on our coast, is ascribed by Da Costa to Dr. Richard Pultney, F. R. S. of Blandford in Dorfetshire ; he found it on an *ascidia* at Weymouth in that county. It has also been met with in Greenland, Iceland, and Norway; and is likewise noted as a native of the Southern Ocean *.

The shell found on the British coast is very small, brittle, and ſemi-transparent. The outside is of a brownish or rosy colour, tinged with green. The inside smooth, glossy, and somewhat pearly.

* *Gmelin.*—Probably this variety found in the South Seas is that which *Da Costa* notices in his description of Mytilus discors. " All that *Linné* has *seen,*" (of Mytilus discors) " as well as all those found on our *coasts,* are very small, thin, and delicate; but a kind no wise different, except in *size* and *colour,* being larger than a great walnut, and quite brown, was brought from the southern hemisphere by that great and national honor, Capt. Cook, the circumnavigator, in the late expedition for the discoveries of new countries. These also were entirely *unknown* to all our collectors; and, as they only differ in *size, thickneſs,* and *colour,* but are exactly the same in structure, way of life, and other particulars as theſe of our coasts, is it a *distinct species* or *variety* only?"— As a figure of this very analogous kind may be acceptable, it is introduced in the annexed Plate at fig. 2.

26

PLATE XXVI.

ANOMIA EPHIPPIUM.

LARGER OR ONION-PEEL ANOMIA.

GENERIC CHARACTER.

Bivalve. Valves unequal: one gibbous towards the beak, the other flat, and perforated near the hinge.

SPECIFIC CHARACTER

AND

SYNONYMS.

Roundish: pellucid, much wrinkled. Flat valve perforated.

ANOMIA EPHIPPIUM: testa suborbiculata rugosa plicata planiore perforata. *Gmel. Lin. Syst. Nat. Conch. p.* 3340. *sp.* 3.

Anomia. Subrotunda plicata pellucida levis, valva planiore perforata. Tunica cepæ. *Da Costa. Br. Conch. p.* 165. *tab.* 11. *fig.* 3.

Huitre. Pelure d'oignon. *Angenv. Conch.* 2. *p.* 316. *tab.* 22. *fig. C.* 11. *p.* 277. *tab.* 19. *fig. C.*

The perforated Oyster. *Petiv. Mus. p.* 85. *No.* 823.

Anomia Ephippium, larger. *Penn. Brit. Zool. No.* 70. *tab.* 62.

PLATE XXVI.

The Anomia Ephippium is frequently found on the common oyster, to the shell of which it adheres by means of a strong tendinous ligature, which passes through the perforation of the upper valve. This Shell is of an irregular form ; the outside rugged and filmy ; the inside smooth, pearly, and glowing with a variety of elegant tints. In different specimens the colours vary considerably, some being of a rich purple, others pale red, brown, or deep yellow, and all with a silvery hue.

27

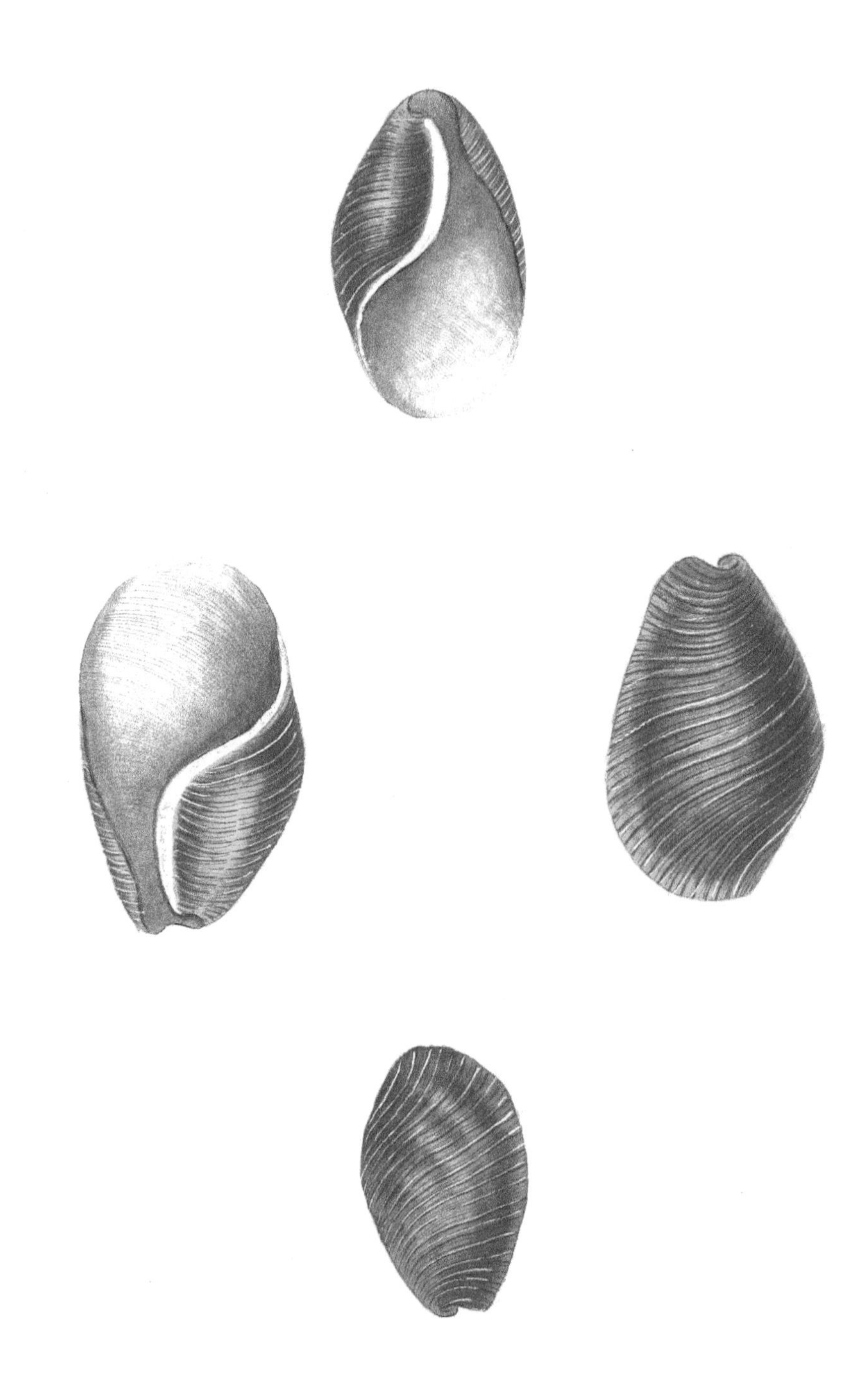

PLATE XXVII.

BULLA LIGNARIA.

WOOD DIPPER.

GENERIC CHARACTER.

Shell sub-oval. Aperture oblong, very patulous, and smooth or even. One end rather convoluted.

SPECIFIC CHARACTER AND SYNONYMS.

Oblong, oval, narrow towards one end, and somewhat umbilicated, (or rather convoluted). Striated transversely.

BULLA LIGNARIA: testa obovata oblongiuscula transverse striata, vertice subumbilicato. *Linn. Syst. Nat.*

BULLA LIGNARIA major, leviter et dense transverse striata. *Da Costa. Br. Conch. p.* 26. *sp.* 14. *tab.* 1. *fig.* 9.

Concha veneris major, leviter et dense striata. *List. H. Conch. tab.* 714. *fig.* 71.

Bulla lignaria. Wood. *Penn. Brit. Zool. No.* 83. *tab.* 70. *fig.* 83.

Oublie, ou papier roulé, tonne a bouche entiere. *D'Avila. Cab. p.* 206, *No.* 387.

This species is not very common. It is found on the coasts of Cornwall, Devonshire, and Dorsetshire, and also on several of the coasts of Ireland.

PLATE XXVII.

The length is generally from one inch and an half to two inches; the shell is brittle and without gloss, of a light, brownish colour, wrought transversely, with fine striæ, and many narrow whitish veins. Its Latin and English names are derived from its supposed resemblance to a piece of veined wood.

This Shell is very open; its animal a slug.

28

PLATE XXVIII.

TURBO CLATHRATUS.

BARRED or FALSE WENTLETRAP.

GENERIC CHARACTER.

Animal Limax. Univalve. spiral, or of a taper form. Aperture somewhat compressed, orbicular, entire.

SPECIFIC CHARACTER AND SYNONYMS.

Shell taper, without umbilicus. Spires swelled and separated by a deep channel. Several regular elevated ribs or ridges extend in a longitudinal direction from the aperture to the apex.

TURBO CLATHRATUS: testa turrita ex umbilicata: anfractibus contignis lævibus. *Gmel. Lin. Syst. Nat. Conch. p.* 3603. *sp.* 63.—*Faun. Suec.* 2170.

Strombiformis minor albus aut pullo variegatus, costis longitudinalibus elatis eleganter distinctus. CLATHRATUS. *Da Costa, Br. Conch. p.* 115. *sp.* 67. *tab.* 7. *fig.* 11.

Cochlea variegata, striis raris admodum eminentibus exasperatae. *List. H. Conch. tab.* 588. *fig.* 51.

Turbo Clathratus. Barred Wentletrap. *Penn. Br. Zool. No.* 111. *tab.* 81. *fig.* 111. 111. *A.*

Fausse scalata. *D'Avila, p.* 221. *No.* 427.

PLATE XXVIII.

This is one of the most singular species that is found on the British coasts. It is very analagous to the famous Scalaris or Wentletrap of the East Indies, which bears such a high price amongst Conchologists; and from this analogy it is called the False Wentletrap. Its length is about an inch, and sometimes two inches or even more. The mouth is perfectly round, and bordered with a thick ring ; from this ring arise several distinct equi-distant prominent ridges, generally eight in number, which extend the whole length of the shell in an obliquely longitudinal direction. These ridges appear the more remarkable and prominent, as the spires are very convex or swelled, and separated from each other by a deep spiral channel. The colour of most specimens is milk white, but is sometimes obscured with brown, or marked transversely with distinct circles of ferruginous interrupted lines.

Turbo Clathratus is found on several of the British coasts.

29

1

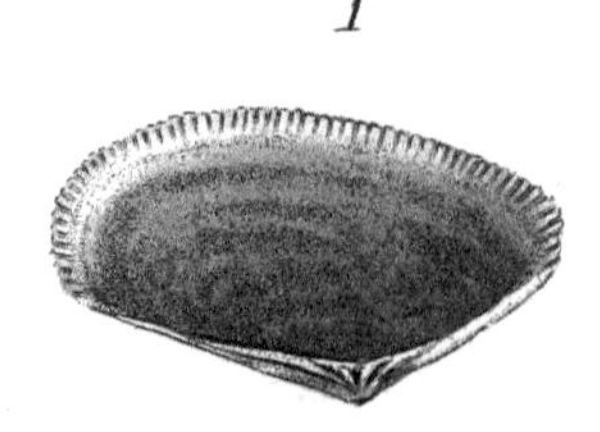

2

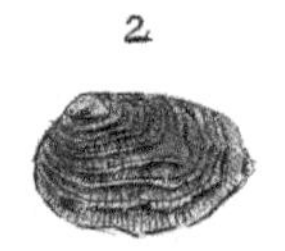

2

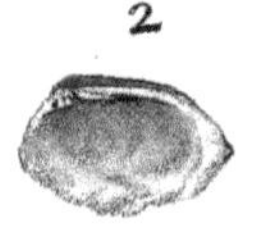

1

PLATE XXIX.

FIG. I.

DONAX TRUNCULUS.

RIBBAND.

GENERIC CHARACTER.

Bivalve. Frontal margin very blunt.

SPECIFIC CHARACTER AND SYNONYMS.

Shell shallow, glossy. Outside fasciated with brown and purple. Inside purple. Margin crenated.

DONAX TRUNCULUS: testa antice lævi intus violacea, marginibus crenatis. *Linn. Syst. Nat.*

Tellina subfusca angustior, inter purpurascens. *List. H. Conch. tab.* 376. *fig.* 217.

Cuneus angustior lævis subfuscus vittis purpurascentibus fasciatus vittatus. *Da Costa, Br. Conch. p.* 207. *sp.* 41.

Donax trunculus. Yellow. *Penn. Br. Zool. No.* 45. *tab.* 55. *fig.* 45.

This pretty species is about one inch and a half in length. It is found on the coasts of Essex, Sussex and Cornwall, and also on those of Wales, Scotland and Ireland.

PLATE XXIX.

FIG. II.

DONAX IRUS.

FOLIATED PURR.

SPECIFIC CHARACTER.

Oval. Outside rugged or wrinkled transversely with numerous raised membranous waved laminæ or foliations.

DONAX IRUS: testa ovali, rugis membranaceis erectis striatis cincta.

Cuneus parvus albescens, rugis foliaceis et membranaceis erectis transversim cinctus. Foliatus. *Da Costa. Brit. Conch. p.* 204.—*Sp.* 39. *tab.* 15. *fig.* 6.

Found in abundance in Cornwall buried in the sands, and not uncommon on the shores of Dorsetshire.

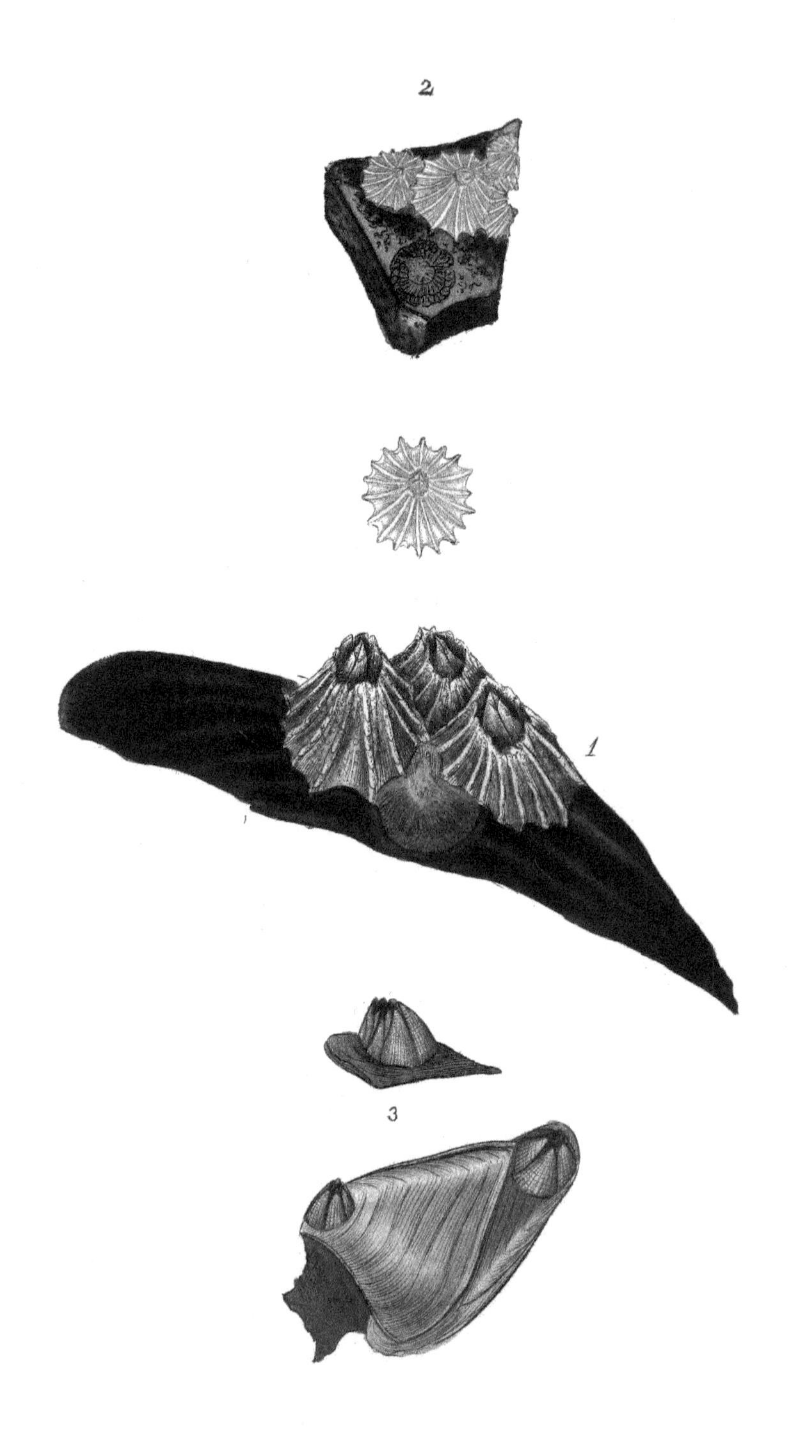
2
1
3

PLATE XXX.

FIG. I.

LEPAS BALANUS.

RIDGED ACORN SHELL.

GENERIC CHARACTER.

Animal Triton. Shell of many unequal valves; affixed by a stem.

SPECIFIC CHARACTER

AND

SYNONYMS.

Shell conic, deeply furrowed, or wrought with prominent longitudinal ridges. Operculum ſharp pointed.

LEPAS BALANUS: testa conica sulcata fixa, operculis acumunatis. *Linn. Syst. Nat. p.* 1107.

Balanus majusculus valvis porcatis. Porcatus. *Da Costa, Br. Conch. p.* 249. *sp.* 69.

Frequent on the British coasts, adhering to rocks, shells, &c.—It is a large and ſtrong species, being seldom less than the size of a filbert, of a conic form and rugged appearance, and is wrought with very prominent longitudinal ridges.

PLATE XXX.

FIG. II.

LEPAS COSTATA.

RIBBED ACORN SHELL.

SPECIFIC CHARACTER.

Shell somewhat conic. Ribs equidiſtant and diverging from the aperture. Operculum sharp pointed.

LEPAS COSTATA: testa subconica operculis acutis: valvulis costatis.

This curious and rare species, which has not been hitherto described or figured, was found by the late T. Adams, Esq. of Pembroke, adhering to pieces of broken rock, and is in the poſſeſſion of the Rev. T. Rackett, of Spetisbury, Dorset, to whose liberality we indebted for figures of this, and several other Britiſh ſhells not included in our own collection.

FIG. III.

LEPAS CONOIDES.

SPECIFIC CHARACTER.

Conic. Smooth, valves pointed at the apex: aperture very small.
LEPAS CONOIDES: testa conica lævi valvulis acuminatis, apertura angustiſſima.

Found by Mr. Bryer of Weymouth, affixed to the Lepas anatifera.

PLATE XXXI.

MUREX DESPECTUS.

THE LARGE or DESPISED WHELK.

GENERIC CHARACTER.

Spiral, rough. The aperture ending in a strait and somewhat produced gutter or canaliculation.

SPECIFIC CHARACTER
AND
SYNONYMS.

Mouth wide, oval and somewhat elongated and cancellated at the upper end. Spires eight.

MUREX DESPECTUS: testa patulo subcaudata oblonga anfractibus octo. *Gmel. Linn. Syst. Nat. p.* 3547.

Buccinum album læve, maximum, septem minimum spirarum. *List. H. An. Angl. p.* 155. *tit.* 1. *tab.* 3. *fig.* 1.

Buccinum rostratum majus crassum, orbibus paululum pulvinatis. *List. H. Conch. tab.* 913. *fig.* 4.

Murex Despectus. Despised. *Penn. Br. Zool. t.* 78. *fig.* 93.

Buccinum canaliculatum magnum crassum striatum album. MAGNUM. *Da Costa tab.* 6. *fig.* 4. *p.* 120.

This is the largest of the turbinated univalves found in the British seas. It inhabits deep water, and is said to be a common shell on the

K 2

Essex, Sussex, and many other of the Englifh shores as well as in Scotland, the Orkneys and many of the Irish shores also. On the Dorset coast it is rare.

It is frequently drawn up with oysters, and is sometimes eaten; but as it is coarse food, it more commonly furnifhes bait to fifhermen.

The largest shells of this sort, found in our seas, sometimes exceeds five inches in length; it is a strong, thick, and heavy shell; of a whitifh colour on the outfide; within of a most lovely yellow, inclining to orange, smooth, and very gloffy.

32

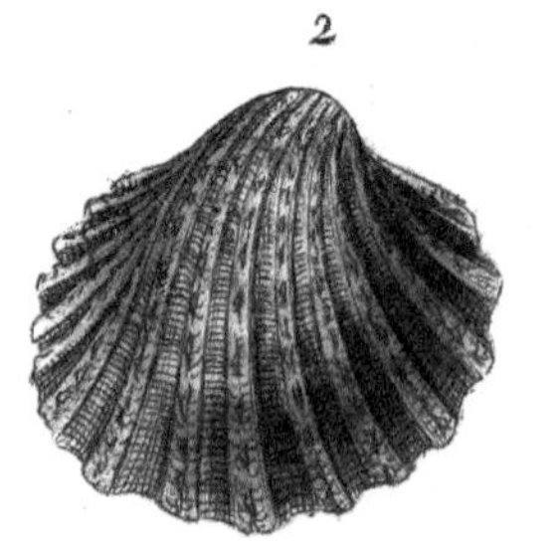

PLATE XXXII.

FIG. I.

CARDIUM MEDIUM.

PIGEON'S HEART COCKLE.

GENERIC CHARACTER.

Two teeth near the beak ; and another remote one on each ſide of the shell.

SPECIFIC CHARACTER
AND
SYNONYMS.

Shell somewhat heart-shaped, and furrowed longitudinally, retuse on one ſide.

CARDIUM MEDIUM : testa subcordata, antice retusa longitudinaliter striato sulcata. *Linn. Syst. Nat. n.* 77. *p.* 1122. *List. Conch. t.* 316. *fig.* 152. *Gualt. t.* 83. *f. b. Chemn. Conch. t.* 16. *fig.* 162.—165.

This shell has not hitherto been noticed as of Engliſh growth. Our specimen, which differs in no respect from those found in the Mediterranean sea, was found near Hartlepoole, on the coast of Durham.

PLATE XXXII.

FIG. II.

CARDIUM CILIARE.

FRINGED COCKLE.

SPECIFIC CHARACTER

AND

SYNONYMS.

Shell roundish, inclining to heart shape. Ribs longitudinal, triangular, and beset along the ridges with thin spines.

CARDIUM CILIARE: testa subcordata, sulcis elevatis triquetris: extimis aculeato ciliatis. *Linn. Syst. Nat. p.* 1122. 80.

Pectunculus albus exiguus, muricibus insigniter exasperatus. *Wallace Orkn. p.* 44.

Pectunculus minimus triquetrus Essexiensis. *Petiv. Gaz. tab.* 93. *fig.* 11.

Cardium parvum tenue, costis triquetris aculeatis. Parvum. *Da Costa Brit. Conch. p.* 177. 17.

Pennant describes this species as having eighteen ribs, and Da Costa about fifteen; we have specimens that agree, in this respect, with the descriptions of both authors. The shell figured by the firſt is the size of a hazel nut; the latter says, he has never seen it larger than a nutmeg: a worn shell, with the habit of this species, that has been found since, is full twice that size.

This delicate shell is found on several of our coasts, as Cornwall, Dorsetshire, and Devonshire; alſo in the Orkneys.

PLATE XXXII.

FIG. III. III.

CARDIUM PYGMÆUM.

PYGMY COCKLE.

SPECIFIC CHARACTER
AND
SYNONYMS.

Nearly heart-shaped, somewhat angulated, furrows imbricated or beset with recurvated scales.

CARDIUM PYGMÆUM: testa subcordata, subangulata, sulcis recurvato imbricatis.

Cardium exiguum. *Gmel. Linn. Syst. Nat. p.* 3255. *sp.* 37?
List. Conch. t. 317. *f.* 154.
Testacea minuta rariora *t.* 3. *f.* 83.

Size of a large currant, of a reddiſh brown, or sometimes whitish colour.

Found in Kent, and at Falmouth, in Cornwall

33

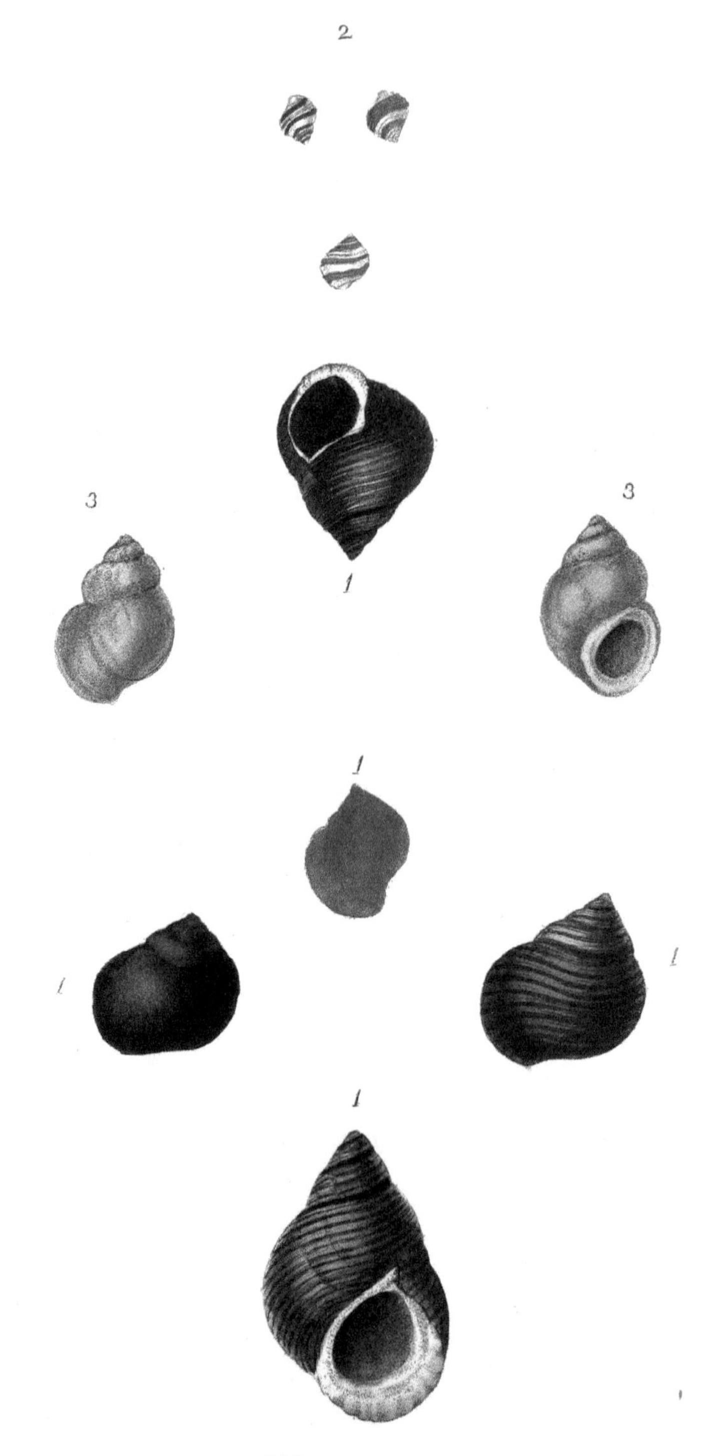

PLATE XXXIII.

FIG. I. II.

TURBO LITTOREUS.

COMMON PERIWINKLE.

GENERIC CHARACTER.

Animal Limax. Univalve, spiral, or of a taper form. Aperture somewhat compressed, orbicular, entire.

SPECIFIC CHARACTER AND SYNONYMS.

Shell with five spires: the firft much swelled, the turban tapering and sharp pointed; striated spirally. Lip thin, and much spread on the pillar.

TURBO LITTOREUS: testa subovate acuta striata, margine columnari plano. *Linn. Syst. Nat. p.* 1232. *No.* 607. *Fn. Sv.* 2. 2169.

Turbo Littoreus, Periwinkle. *Penn. Br. Zool. No.* 109. *tab.* 81. *fig.* 109.

Turbo-pyramidalis crussus fuscus, striis crebris prædítus. Littoreus. *Da Costa, Br. Conch. p.* 98. 55.

It is needless to enter into detail on a species so well known as the Common Periwinkle; yet it may be proper to observe, these shells, in many instances, vary considerably. The young shells are reddish, whitish, yellowish, or brown, of various hues, sometimes uniform, or without any markings; at others, girdled spirally with darker colours. The adults also are sometimes bright red, orange, chesnut, or whitish, or olive, with, and sometimes without, the spiral lineations. They vary no less in size than in colours; and those of the Orkneys, in particular, are quadruple, the size of those on the Irish or English coast.

It is said, the name Periwinkle is a corruption of *Petty Winkle*, or small Winkle, or Whelk.

Fig. I. I., &c. Adult vanities of Turbo Littoreus. Fig. II. II. the young shells.

FIG. III. III.

TURBO RUDIS.

THICK-LIPPED PERIWINKLE.

SPECIFIC CHARACTER.

Shell somewhat tapering, without umbilicus. Volutions of the spires, or turban swelled. Lip thick, and glossy within.

This shell has been kindly communicated by Dr. Maton, as a new species. It is noticed in the first volume of that gentleman's Observations on the WESTERN COUNTIES; but has not, we believe, been hitherto figured. It was discovered on the banks of the Tamar, in Devonshire, near Bere-Alston, and is thus described:

PLATE XXXIII.

" In the mud appeared a species of *Turbo*, which, though very similar to *T. Littoreus* (the common Periwinkle), has some characters that seem to authorize its being considered as a different shell. The *anfractus* are much more swollen, as it were, than in the above species; the spire is more depressed; and, besides, there is no appearance of *striæ*, either transversely or longitudinally. This shell has a sort of distorted or rude *contour*, that may, perhaps, entitle it to the appellation of *T. Rudis*. Its colour is greenish." *Page* 277, *vol.* 1.

34

PLATE XXXIV.

OSTREA PUSIO.

DISTORTED.

GENERIC CHARACTER.

Animal a Tethys. Shell bivalve, unequal. The hinge without a tooth, having a small oval cavity.

SPECIFIC CHARACTER AND SYNONYMS.

Eared. With about forty longitudinal rays. Shell irregular, or diſtorted.

Concha testa aurita, striis circiter quadraginta. *Linn. F. Suec.* 1. *p.* 384. *No.* 1345.

Pecten minimus angustior, inæqualis fere et asper, sinu ad cardinem cylindraceo, ceberrimis minutissimisque striis donatus. *List. Hist. Angl. p.* 186. *tit.* 31. *tab.* 5. *fig.* 31.

Pecten Pusio. Writhed. *Penn. Br. Zool. No.* 65. *tab.* 61. *fig.* 65.

Pecten minor alba, contusa proteiformis. *Petiv. Gazoph. tab.* 94. *fig.* 2.

Twisted Pectines *of Stroma. Wallace, Orkn. p.* 43, 44

Pecten parvus inæqualibus, informis, striatus. DISTORTUS, DISTORTED. *Da Costa, Br. Conch. p.* 148. *tab.* 10. *fig.* 3. 6.

PLATE XXXIV.

Dissimilar as the several figures in this plate may appear, they are merely accidental varieties in size, growth, and colour of an individual species. These shells are generally about an inch, or more, in length, and of a somewhat globose shape; but so extremely irregular and distorted, that it can with difficulty be defined. Dr. Wallace calls these shells the twisted pectines of Stroma, a little island that lies in the Pightland Frith; he found them on some parts of the Orkneys, where he observed extraordinary cross and strong tides. The irregular form of these pectines rather surprised him. He adds, "I cannot think the odd strange tumbling the tides make there, can contribute any thing to that frame; yet, after all, I never see them in any other place."— These shells are now found on several of the English shores, as Yorkshire, Scarborough, Mouth of the River Tees, and Dorsetshire; and are also frequent in the fossil state, in the chalk pits of Kent and Surry. Both valves are convex, and much diftorted, but the under one is usually the most irregular; the ribs are numerous, close set, longitudinal, and prominent. The inside of the lower valve is smooth and white, when alive, and that of the upper has a pearly gloss. The outside is generally of a dingy white, or yellowish cast; sometimes pale violet, or russety; or white mottled, and varied with brown, or brilliant red.

The ears of this shell are large, and nearly equal, but are often so distorted as to appear much otherwise. It is proper to observe, that though the irregular form this shell assumes may be attributed to some injury it has sustained in its growth, every shell of this species is conftantly found with the same diftorted appearance. This circumftance leaves no reason to doubt that such diftortions are characteriftic of this extraordinary and peculiar species.

35

PLATE XXXV.

MUREX ERINACEUS.

GENERIC CHARACTER.

Spiral, rough. The aperture ending in a strait and somewhat produced gutter or canaliculation.

SPECIFIC CHARACTER.

AND

SYNONYMS.

Shell subangular, rugose, or covered entirely with raised scales or points. Spires six.

MUREX ERINACEUS: testa multifarium subfrondoso-spinosa, spiræ anfractibus retuso coronatis, cauda abbreviata. *Linn. Syst. Nat.* 526. *p.* 1216.— *Gmel. Linn. Syst. Nat. p.* 3530.

Buccinum majus canaliculatum, rostratum, ore labioso, fimbriatum, umblicatum, ore angusto, oblongo, rugosum, costulatum, striis eminentibus reticulatim exasperatum, albidum. *Gualt.* 1. *Conch. Tab.* 49. *fig.* H.

MUREX ERINACEUS. Urchin. *Penn. Br. Zool. No.* 95. *tab.* 76. *fig.* 95. *Seba. Muf.* 3. *t.* 49. *fig.* 78, 79.— *Martin. Conch.* 3. *t.* 110. *f.* 1026—8.

Buccinum longirostrum medium subangulatum, porcis spiralibus distinctum. Porcatum. *Da Costa. tab.* 8. *fig.* 7. 7. *p.* 133.

PLATE XXXV.

Found on the coast of Cornwall, and Dorsetſhire, and also on that of Hilbree iſland in Cheshire.

36

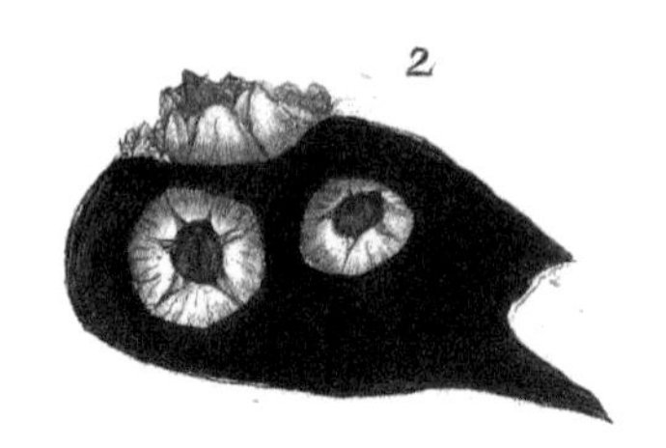

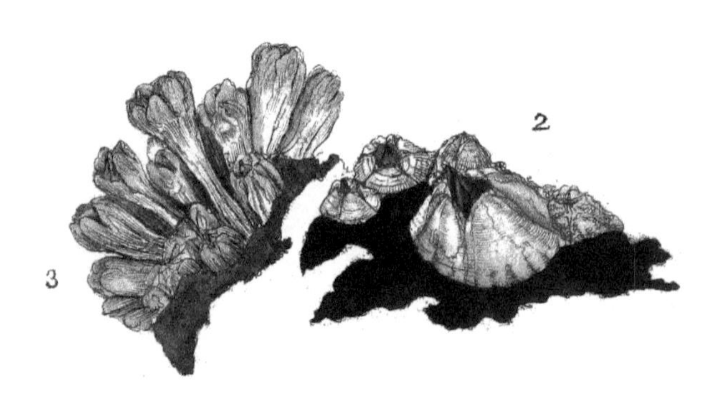

PLATE XXXVI.

FIG. I.

LEPAS INTERTEXTA.

GENERIC CHARACTER.

Animal Triton. Shell of many unequal valves: affixed by a stem.

SPECIFIC CHARACTER

AND

SYNONYMS.

Shell rather depressed and ribbed obliquely.

LEPAS INTERTEXTA: testa subdepressa oblique costata.
Lepas striata. *Penn. Br. Zool. t.* 38. *f.* 7.
Walker teſt. min. rar. f. 87.

This rare ſpecies is the *Lepas intertexta* of the Portland Museum; it was fiſhed up at Weymouth, adhering to a valve of the Ostrea ſub-rufus.

The shells of this genus are in general very complex in structure, the present is particularly so. Several shells of the natural size is represented on the orange space of the Ostrea, Fig. I.—Three figures are added to exhibit their magnified appearance; to distinguish these, the space is coloured green.

PLATE XXXVI.

FIG. II. III.

LEPAS BALANOIDES.

COMMON ACORN SHELL.

SPECIFIC CHARACTER

AND

SYNONYMS.

Shell conic truncated, of six valves. Operculum bifid.

LEPAS BALANOIDES: testa conica truncata; operculo obtuso *Linn. Faun. Suec.* 1. *p.* 385. *No.* 1348 2. *No.* 2123.

Balanus vulgaris parvus conicus é senis laminis compositus, vertice operculo bifido rhomboide occluso. *Da Br. Conch. p.* 248. *sp.* 68. *tab.* 17, *fig.* 7.

Balanus cinereus, velut é senis laminis striatis compositus, ipso vertice altera testa, bifida, rhomboide occluso. Balani parva species. *List. H. An. Angl. p.* 196. *tit.* 41. *tab.* 5. *fig.* 41.

These shells are found in the greatest abundance on all the British shores, adhering to *rocks*, *shells*, *&c. &c.*

Da Costa says, this species, when not affixed on flat, but uneven surfaces, sometimes, but rarely, extend down into a pretty long rugged tubular stalk or root. This variety is noted by Pennant, and an extraordinary, but mutilated specimen of it, is shewn at fig. 3.

INDEX.

VOL. I.

LINNÆAN ARRANGEMENT.

MULTIVALVIÆ.

BIVALVIA. CONCHÆ.

L 2

I N D E X.

UNIVALVIA.

INDEX TO VOL. I.

ACCORDING TO

HISTORIA NATURALIS TESTACEORUM BRITANNIÆ OF DA COSTA.

GENUS I. PATELLA, LIMPET FLITHER OR PAP SHELL.

* MARINÆ. SEA

GENUS 2. HALIOTIS, EAR SHELL.

GENUS 3. SERPULA. WORM SHELL.

INDEX.

PART II.

UNIVALVIA INVOLUTA.

GENUS 5. BULLA. DIPPER.

* * MARINÆ. SEA.

PART III.

UNIVALVIA TURBINATA.

TROCHUS TOP SHELL.

* MARINÆ. SEA.

COCHLEÆ, OR SNAILS.

GENUS 8. NERITA, NERIT.

* FLUVIATILES. RIVER.

MARINÆ SEA.

GENUS 9. HELIX.

* TERRESTRES. LAND.

HELIX.

I N D E X.

*** MARINÆ. SEA.

		Plate	Fig.
Cochlea catena. Chain	- - - -	20.	1. 1.

GENUS 11. TURBO.

* TERRESTRES. LAND.

		Plate	Fig.
Turbo fasciatus. Fasciated	- - - -	18	1. 1.

*** MARINÆ. SEA.

		Plate	Fig.
Turbo Littoreus, periwinkle	- - - -	33	1. 2.
Turbo pictus, painted	- - - -	2	2, 3, 4, 5, 6
Turbo cancellatus, latticed	- - - -	[illegible]	1. 1.

GENUS 13. STROMBIFORMIS. NEEDLE SNAIL.

		Plate	Fig.
Strombiformis terebra, auger	- - - -	22	2.
Strombiformis cinctus, girdled	- - - -	22	1. 1.
Strombiformis clathratus, barred or false wentletrap	-	28	

BUCCINA, WILKS, OR WHELKS.

GENUS 13. BUCCINA CANALICULATA.

GUTTERED WHELKS.

* MARINÆ. SEA.

		Plate	Fig.
Buccinum magnum, large	- - - -	31	
Purpuro-Buccinum, purple whelke	- - -	11	

I N D E X.

MURICES, ROCKS.

GENUS 16. APORRHAIS.

* MARINÆ. SEA.

ORDER 2.

BIVALVES.

GENUS 1. PECTEN. ESCALLOP.

GENUS 3. ANOMIA.

* MARINÆ. SEA.

DIVISION 2.

GENUS 6. CARDIUM. HEART COCKLE.

* * MARINÆ. SEA.

INDEX.

GENUS 7. PECTUNCULUS. COCKLE.

* MARINÆ. SEA.

	Plate.	Fig.
Pectunculus glaber, smooth - - -	17	

GENUS 9: CUNEUS. PURR.

* MARINÆ. SEA.

	Plate.	Fig.
Cuneus fasciatus, fasciated		
Cuneus foliatus, foliated - - - -	29	2. 2.
Cuneus truncatus, truncated - - -	24	
Cuneus vittatus, ribband - - - -	29	1. 1.

GENUS 11. MYTILUS MUSCLE.

* MARINÆ. SEA.

	Plate.	Fig.
Mytilus Modiolus, great - - - -	23	
Mytilus difcors, divided - - - -	25	

GENUS 15. PINNA. SEA HAM OR WING.

* MARINÆ. SEA.

	Plate.	Fig.
Pinna muricata, thorny - - - -	10	

PART IV.

MULTIVALVES.

GENUS 18.

	Plate.	Fig.
Balanus vulgaris, common - - - -	36	2. 3.
Balanus porcatus, ridged - - - -	30	1.
Balanus anatiferus barnacle - - - -	7	

ALPHABETICAL INDEX TO VOL. I.

	Plate.	Fig.
ACULEATUM, Cardium, Spiked Cockle	6	
Anatifera, Lepas, Barnacle	7	
Balanoides, Lepas	36	2.
Balanus, Lepas	30	1. 1.
Bimaculata Tellina, Double Spot Tellen	19	1. 1.
Chione, Venus	17	
Ciliare, Cardium	32	2.
Cimex, Turbo, Latticed Whelk	2	1. 1.
Cinctus, Turbo, girdled	22	1. 1.
Clathratus, Turbo, Falſe Wentletrap	28	
Conoides, Lepas, Conio Acornshell	30	3.
Conulus, Trochus, Conule	8	2. 3.
Costata, Lepas, Ribbed Acorn-shell	30	2.
Despectus Murex, Deſpiſed Whelk	31	
Discors, Mytilus, divided	25	1.
Erinaceus, Murex	35	
Fasciatus, Turbo, fasciated	18	1. 1.
Fissura, Patella, Slip Limpet	3	
Fluviatilis, Nerita, River Nerit	16	2. 2.
Glaucina, Nerita, Chain Nerit	20	1. 1.
Hungarica, Patella, Large Fool's Cap, Limpet	21	1. 1.
Intertexta Lepas, Striated Acorn Shell	36	1.
Irius Tellina, Foliated Purr	29	2. 2.
Lapillus, Buccinum, Massy, or Purple Whelk	11	
Lignaria Bulla	27	
Lineatum, Buccinum, lineated	15	
Littoralis, Nerita	20	2. 2.
Littoreus, Turbo	33	1. 1. 2.
Magus, Trochus, Tuberculated Top Shell	8	1.
Medium Cardium, Pigeon's Heart Cockle	32	1.
Modiolus Mytilus	23	
Muricata Pinna, Thorny Wing, or Sea Ham	10	

INDEX.

Nemoralis, Helix, Girdled Snail - - - - 13

Obsoletus, Pecten - - - - - 1 2. 2.

Pallidulus, Nerita, Pale Nerit - - - - 16 1:

Pellucida, Patella, Blue Rayed Limpet - - - 3 1. 1.

Pes Pelecani, Strombus, Corvorant's Foot - - - 4

Pullus Turbo, Painted Whelk - - - - 2 2. 3. 4. 5.

Pusio, Ostrea, distorted - - - - 34

Pygmeum, Cardium, Small Cockle - - - 32 3.

Reticulata, Patella, Reticulated Mask Limpet - - 21 3.

Rudis, Turbo, Thicklipped - - - - 33 3. 3.

Subrufus, Ostrea - - - - - 12

Spirorbis, Serpula, Wrackfpangle - - - - 9

Tenuis, Tellina, Thin Tellen - - - - 19 2. 2.

Terebra, Turbo - - - - - 22 2. 2.

Trunc-culus, Tellina - - - - - 29 1. 1.

Tuberculata, Haliotis, Tuberculated Sea Ear - - 5

Varia Ostrea, Variegated, or One-eared Scallop - - 1

Vulgata, Patella, Common Limpet - - - 14

THE

NATURAL HISTORY

OF

BRITISH SHELLS,

INCLUDING

FIGURES AND DESCRIPTIONS

OF ALL THE

SPECIES HITHERTO DISCOVERED IN GREAT BRITAIN,

SYSTEMATICALLY ARRANGED

IN THE LINNEAN MANNER,

WITH

SCIENTIFIC AND GENERAL OBSERVATIONS ON EACH.

VOL. II.

BY E. DONOVAN, F.L.S.
AUTHOR OF THE NATURAL HISTORIES OF
BRITISH BIRDS, INSECTS, &c. &c.

LONDON:
PRINTED FOR THE AUTHOR,
AND FOR
F. AND C. RIVINGTON, N° 62, ST. PAUL'S CHURCH-YARD.
BY BYE AND LAW, ST. JOHN'S SQUARE, CLERKENWELL.

1800.

37

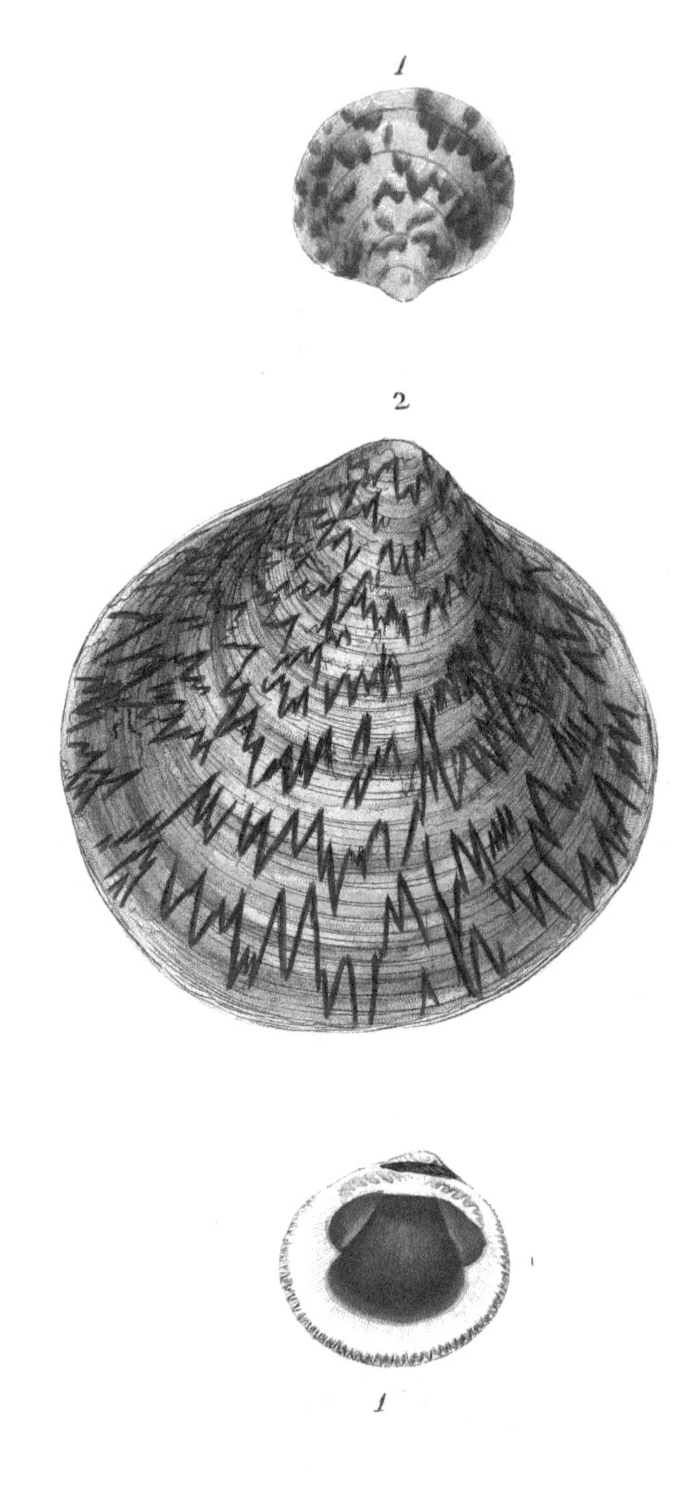

THE

NATURAL HISTORY

OF

BRITISH SHELLS.

PLATE XXXVII.

ARCA GLYCYMERIS.

ORBICULAR ARK.

GENERIC CHARACTER.

Bivalve, valves equal. Teeth of the hinge numerous, and inserted between each other.

SPECIFIC CHARACTER

AND

SYNONYMS.

Orbicular, concave, very finely striated transversely and longitudinally, and variegated with zigzag marks. Margin crenated.

ARCA GLYCYMERIS: testa suborbiculata gibba, substriata, natibus incurvis, margine crenato. *Linn. Syst. Nat. p.* 1143. *No.* 181.

A 2

PLATE XXXVII.

Chama glycemeris, Bellon. Pectunculus ingens variegatus ex rufo. *List. H. Conch. tab.* 247. *fig.* 82.

Concha crassa, lævis, subalbida, luteis maculis radiata, signata, fasciata, et virgulata, intus macula fusca obscurata. *Gualt.* 1. *Conch. tab.* 72. *fig. G.*

Glycymeris cornubiensis crassa marmorata. *Mus. Petiv. p.* 84. *No.* 816.

Bastard, or dog's cockle. *Rutty Dublin, p.* 379.

Arca glycymeris, orbicular, *Penn. Br. Zool. No.* 58. *tab.* 58. *fig.* 58.

Glycymeris. Orbicularis crassa subalbida lineis rufulis sagittæformibus variegata, intus obfuscata margineque crenato. Orbicularis. *Da Costa. Br. Conch. p.* 168, *tab.* 11. *fig.* 22.

This species is found, of a large size, in the Mediterranean sea; those which inhabit the English coast, as Falmouth and Cornwall, rarely exceed the size of the smallest specimen we have represented. It is found likewise on the shores of Guernsey, and the coast of Ireland, where it is called the *dog's cockle*.

38

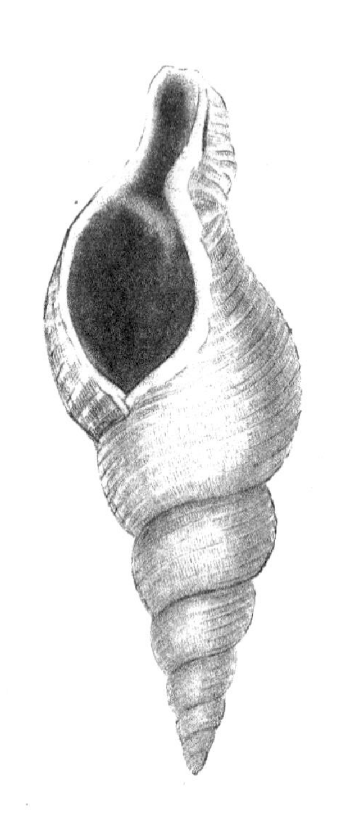

PLATE XXXVIII.

MUREX CORNEUS.

HORNY, or SLENDER WHELK.

GENERIC CHARACTER.

Spiral, rough. The aperture ending in a strait, and somewhat produced gutter, or canaliculation.

SPECIFIC CHARACTER,

AND

SYNONYMS.

Slender, white. Spires eight, swelled. Mouth oblong oval, ending in a produced or lengthened deep twirled gutter.

MUREX CORNEUS: testa oblonga rudi, anfractuum marginibus complanatis, apice tuberculoso, apertura edentula, cauda adscendente. *Linn. Syst. Nat. p.* 1224. *No.* 565.

Buccinum angustius, tenuiter admodum striatum, octo minimum spirarum. *List. H. An. Angl. p.* 157. *tit.* 4. *tab.* 3. *fig.* 4.—*App. H. An. Angl. p.* 15, 16.

Lesser long and smooth whelke, *Dale Harw. p.* 381. *No.* 2.—*Smith Cork, p.* 401. *No.* 7.

Narrow-mouthed whelke, with eight wreaths. *Wallis Northumb p.* 401. *No.* 7.

Murex corneus, Horny. *Penn. Brit. Zool. No.* 99. *tab.* 76. *fig.* 99.

Buccinum canaliculatum medium, angustius, album, striatum, octo spirarum. GRACILE *Da Costa, p.* 124. *sp.* 74. *tab.* 6. *fig.* 5.

A 3

PLATE XXXVIII.

This shell is white, semitransparent, and rather glossy; and when alive is covered with a fine thin brown film, or epidermis, which is striated spirally. It is found on several of the English coasts, as Yorkshire, Northumberland, Essex, &c. and also on the shores of Scotland and Ireland.

39

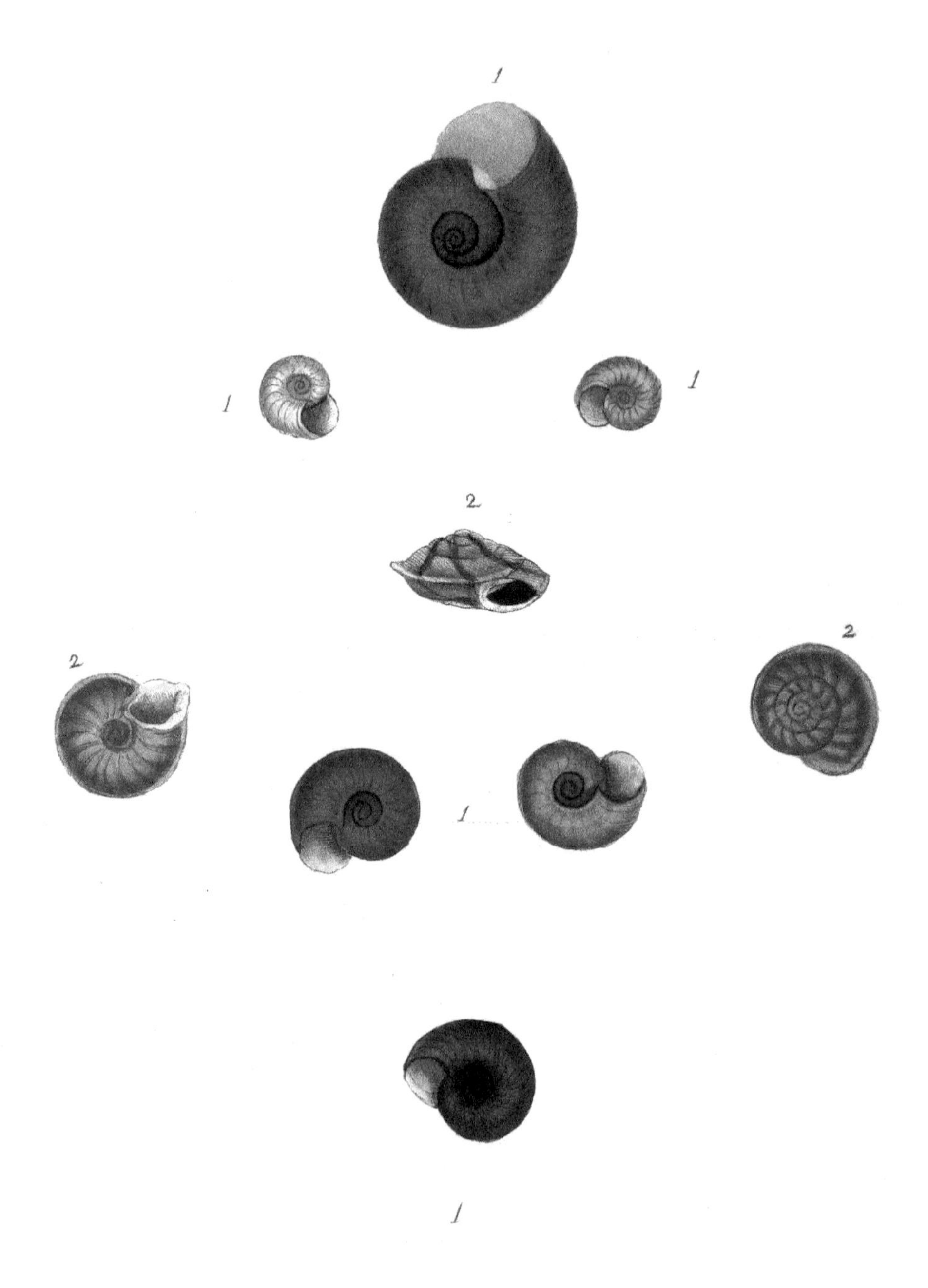

PLATE XXXIX.

FIG. I.

HELIX CORNEA.

RAM'S HORN.

GENERIC CHARACTER.

Aperture of the mouth, contracted, and lunated.

SPECIFIC CHARACTER

AND

SYNONYMS.

Wreaths, four, turned nearly horizontal: rather depressed or concave towards the centre.

HELIX CORNEA: testa supra umbilicata plana nigricante, anfractibus quatuor teretibus. *Lin. Syst. Nat. p.* 1243. *No.* 671. —*F. Suec. I. p.* 373. *No.* 1304. *II. No.* 2179.

Cochlea pulla, ex utraque parte circa umbilicum cava. *List. H. Angl. p.* 143. *tit.* 26. *tab.* 2. *fig.* 26.

Cochlea maxima, compressa, fasciata. *List. H. Conch. tab.* 136. *fig.* 40.—Cochlea pulla quatuor orbium coccum fundens, purpura lacustris. *Id. tab.* 137. *fig.* 41.

The Flat Whirl. Grew. Mus. p. 136.

Planorbis fluviàtilis major vulgaris. *Petiv. Gazoph. tab.* 92. *fig.* 5.

Helix Cornea, Horny. *Penn. Br. Zool. No.* 126. *tab.* 83. *fig.* 126.

A 4

PLATE XXXIX.

Helix fluviatilis depressa major, anfractibus quatuor, ex utraque parte circa umbilicum cava. Cornu arietis. *Tab.* 4. *fig.* 13. DA COSTA *Br. Conch. p.* 60. *tab.* 4. *fig.* 13.

Purpura. S. Cochlea fluviatilis compressa major. *List. Exerc. Anat.* 2. *p.* 59.

The adult shells of this species are from three quarters of an inch, to an inch and a quarter in diameter; the colours various, generally brownish or ashen colour, inclining in some to red, in others to yellow; the young shells are whitish and more transparent.

It is very common in ponds and rivers. The animal is blackish brown, and has two red capillary horns *.

The Helix Nana, or Dwarf of Pennant, fig. 125, is considered by Da Costa and other conchologists, as a young shell of this species.

* " This fish emits a fine *scarlet humour*, if a grain of salt of any kind, or a little pepper or ginger, be put into the mouth of the shell. It emits this fine scarlet humour all the year, especially in April and September. Dr. Lister gives a full account of it. He says, this scarlet humour may be readily got, and in great quantity, if a large parcel of these shells be wrapped up in a cloth bag, sprinkling over it a little salt; then the scarlet liquor will ouze plentifully. The colouring part of this humour immediately subsides, if sprinkled with powdered allum, and the rest of it remains like clear water. The colouring part may be strained through a filtering paper, but the elegance of its colour is lost, and it changes into a dull, unpleasant rusty brown. Moreover, if mixed with vinegar, spirit of wine, deliquated vegetable salts, or common salt dissolved, this elegant scarlet colour perishes in the same manner as when mixed with allum. Neither can this liquor be kept by itself pure and unmixed; for in vain did the doctor strive to preserve it in narrow mouthed bottles or phials, perfectly well closed, and with oil or honey thrown over it. Thus this colour is of so fugitive a nature, that no acid or astringent has been found sufficient, to preserve the elegance of its tint."

" Dr. Lister further recites some observations and experiments he made on this scarlet fluid, to discover whether it was a humour of the body, or to be got by laceration or incision, as blood; a saliva from the throat or stomach; or a particular humour contained in certain vessels or parts; but the nicety and difficulty of the experiments rendered it impossible for him to determine it precisely." Da Costa, page 61, 62.

PLATE XXXIX.

FIG. II.

HELIX LAPICIDA.

ACUTE EDGED.

SPECIFIC CHARACTER

AND

SYNONYMS.

Above and beneath rather convex; back of the wreaths carinated. Deeply umbilicated.

HELIX LAPICIDA: testa carinata umbilicata utrinque convexa, apertura marginata transversali ovata. *Linn. Syst. Nat. p.* 1241. *No.* 656.

Cochlea testa utrinque convexa, subtus perforata, spira acuta apertura ovata transversali. *Linn. Faun. Suec.* 1. *p.* 371. *No.* 1298. 11. *No.* 2174.

Cochlea pulla, sylvatica, spiris in aciem depressis. *List. H. An. Angl. p.* 126. *tit.* 14. *tab.* 2. *fig.* 14.

Cochlea nostras, umbilicata, pulla. *Hist. H. Conch. tab.* 69. *fig.* 68.

Planorbis terrestris Anglicus, umbilico minore, margine acuto. *Mus. Petiv. p.* 69. *No.* 734.

Cochlea terrestris media acie acuta: land cheese shell with a sharp edge. *Petiv. Gaz. tab.* 92. *fig.* 11.

Helix Lapicida. ROCK. *Penn. Br. Zool. No.* 121. *tab.* 83. *fig.* 121.

Cochlea umbilicata, margine in acie acuto depresso, ACUTA. sharp. *Da Costa Br. Conch. p.* 55. *tab.* 4. *fig.* 9. 9.

PLATE XXXIX.

This species is found in several countries of Europe. In Great Britain it seems to be a local and rather uncommon kind. Da Costa says "though found in many parts of England, is not met with in any plenty, but is scarce. I have found them on the rocks, at and near Matlock, in Derbyshire, about Bath, in Somersetshire, also on rocks; in Surrey, Wiltshire, and Hampshire, in the moss on the bodies of large trees, and in woods. Dr. Lister found them on the grass in Lincolnshire; Mr. Petiver, in hedges, between Charlton and Woolwich, in Kent*; Mr. Morton, in hedge-bottoms, in Oakly Parva, in Northamptonshire; and Mr. Wallis, on the rocks in Northumberland: but they are not common or frequent any where." page 56.

* Not uncommon last summer in the woods of Kent.

40

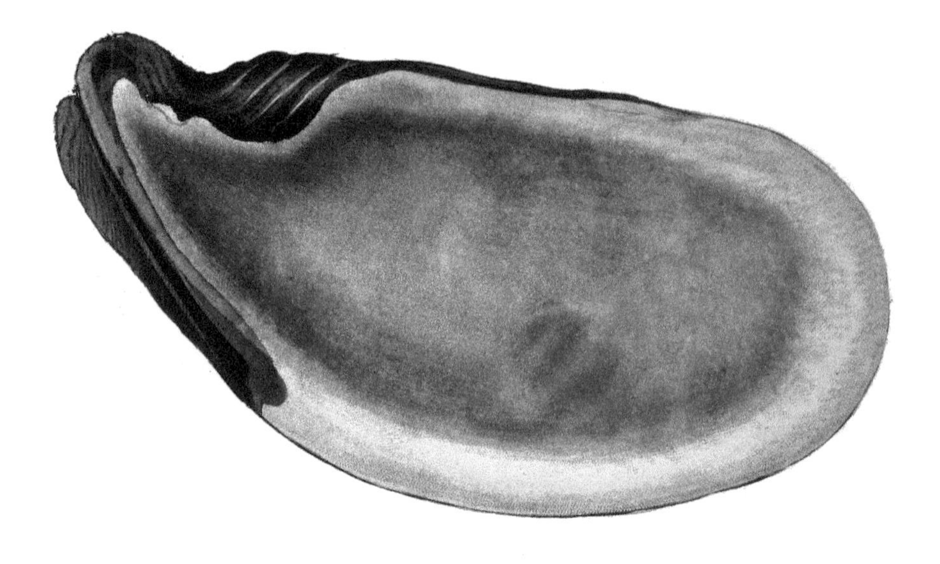

PLATE XL.

MYTILUS UMBILICATUS.

UMBILICATED, OR WRY BEAK MUSCLE.

GENERIC CHARACTER.

The hinge toothless, and consists of a longitudinal furrow.

SPECIFIC CHARACTER AND SYNONYMS.

Hinge much depressed and bent inwards.

MYTILUS UMBILICATUS, umbilicated. *Penn. Br. Zool. sp.* 76. *fig.* 76.

MYTILUS CUROIROSTRATUS. WRY BEAK. *Da Costa Br. Conch. p.* 220. 50.

We are informed by Pennant that this shell was discovered by the Reverend Hugh Davies; that it is a rare and new species, and is sometimes dredged up off Priestholme Island, Anglesea.

It is about half the size of Mytilus Modiolus, and in some respects resembles it; but is distinguished by the very remarkable and peculiar structure of its hinge; the space opposite to it is bent inwards, in a winding manner, into a deep rugged cavity, which when the shells are closed, form a deep hollow, or umbilicus, as if bruised in. On one valve this depression is more deeply inflected inwards than on the other.

41

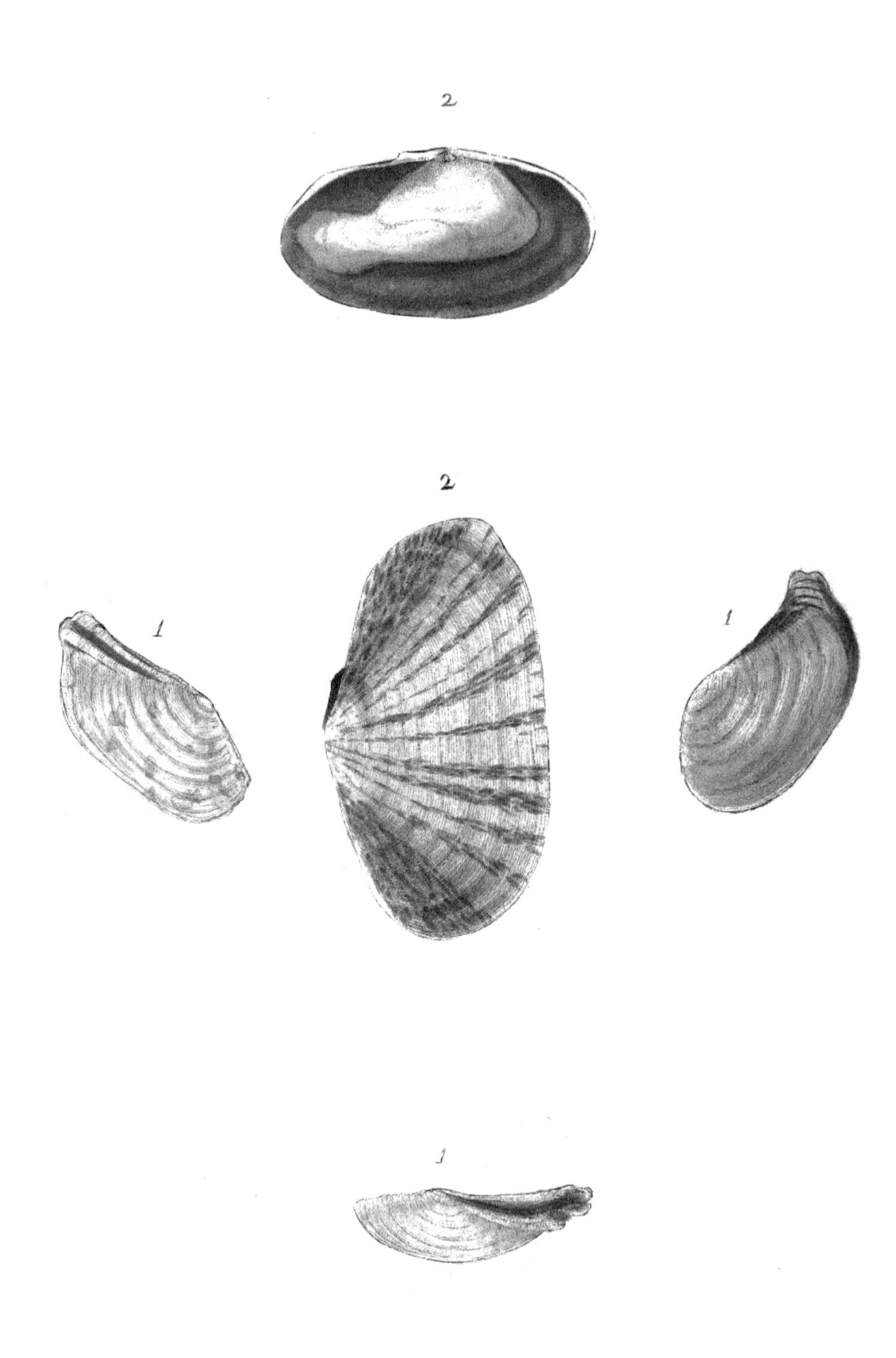

PLATE XLI.

FIG. I.

TELLINA INÆQUIVALVIS.

UNEQUAL-VALVED TELLEN.

GENERIC CHARACTER.

The hinge usually furnished with three teeth; shell generally sloping on one side.

SPECIFIC CHARACTER
AND
SYNONYMS.

Oblong, one side much produced or beaked; upper valve flat, lower very convex.

TELLINA INÆQUIVALVIS: testa oblongo-rostrata, valva altera plana.
Gmel. Lin. Syst. Nat. Conch. 3233. *sp.* 23.

The *Tellina inæquivalvis* is noticed by *Gmelin* as a native of the Mediterranean and Norway seas, but has not been hitherto described as a British shell by any author. It is generally admitted by Conchologists that the species has been discovered on our shores, and William Pilkington, Esq. of Whitehall, has very lately received a specimen of it from the Guernsey coast, which he obligingly favoured us with it to figure and describe.

PLATE XLI.

FIG. II.

TELLINA VARIABILIS.

SPECIFIC CHARACTER.

Shell somewhat oval or oblong, radiated with pale red streaks; a single tooth in the hinge of one valve, which is inserted between two teeth on the other valve, when shut.

SOLEN VESPERTINUS: testa ovali oblonga spadiceo-radiata, cardinis sinistræ valvæ dente solitario duplici alterius inserto. *Gmel. Lin. Syst. Conch. p.* 3228. *sp.* 20.

This is the *Tellina variabilis* of the late Dr. *Solander*, and the *Portland Museum*; Gmelin arranges it amongst the Solens.

Found on the coast of Cornwall and Weymouth, and not noticed by either *Pennant* or *Da Costa* as an English shell.

42.

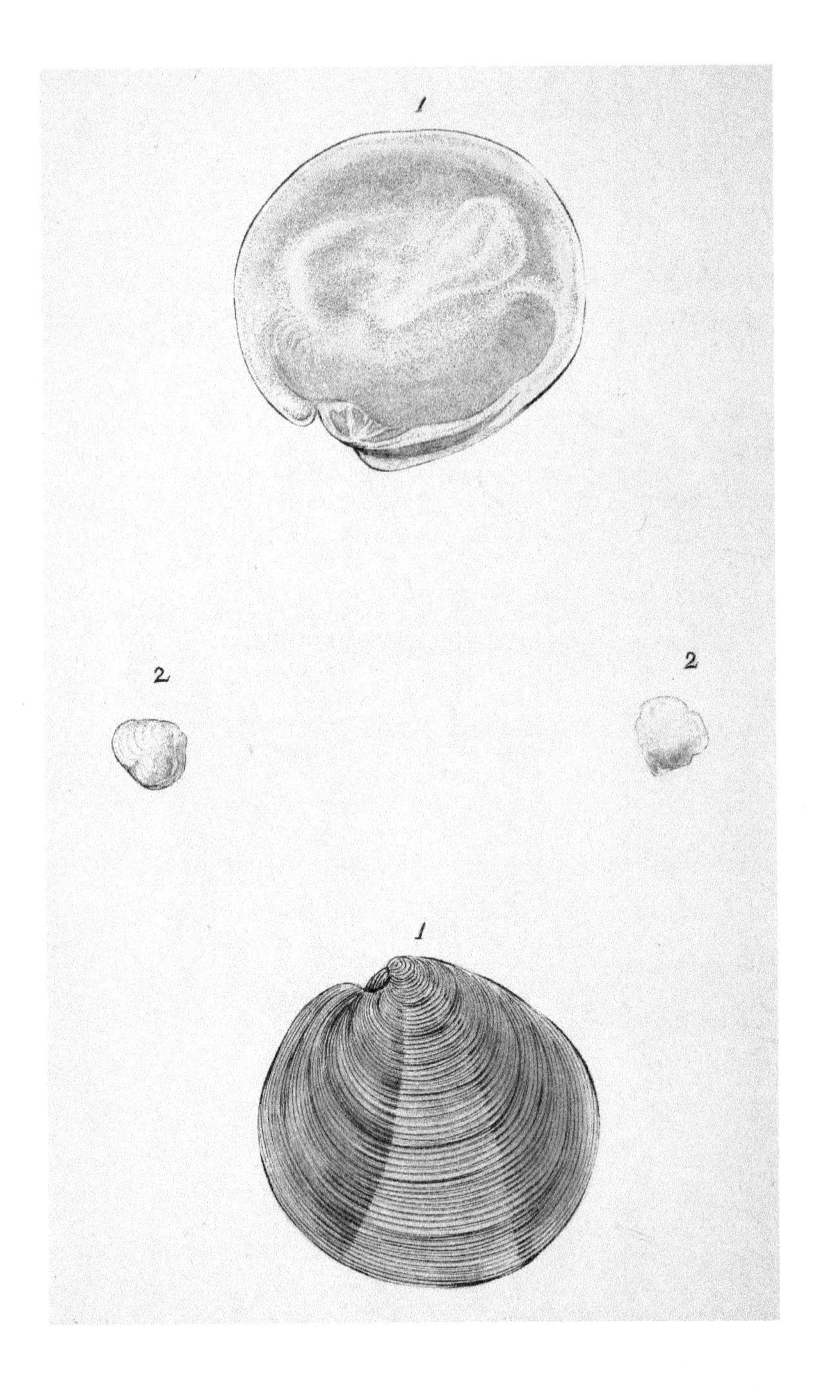

PLATE XLII.

FIG. I.

VENUS EXOLETA.

ANTIQUATED.

GENERIC CHARACTER.

Bivalve. Hinge furnished with three teeth; two near each other, the third divergent from the beaks.

SPECIFIC CHARACTER.

Subrotund. Wrought transversely with numerous regular and minute striæ, margins smooth.

VENUS EXOLETA: testa lentiformi transversim striata pallida, obsolete radiata, ano cordato. *Linn. Syst. Nat. p.* 1134. *No.* 142.

Concha testa subrotunda: striis transversis innumeris, margine lævi. *F. Suec.* 1, *p.* 383, *No.* 1342.

Pectunculus rostro productiore, capillaceis fasciis donatus. *List. H. Conch. tab.* 290. *fig.* 126.—P. dense fasciatus, ex rubro variegatus et undatus. *Tab.* 291. *fig.* 127.—P. crassus, dense fasciatus, leviter ex rufo variegatus. *Tab.* 292. *fig.* 128.—P. subfuscus tenuiter admodum fasciatus. *Tab.* 293. *fig.* 129. P. planus, crassus, exrufo radiatus. *Tab.* 299. *fig.* 136.

Concha marina valvis æqualibus æquilatera, notabiliter umbonata et oblique incurvata, subrotunda, vulgaris, striis densissimis et profundis transversim striata et exasperata, candida leviter ex fusco variegata et radiata. *Gualt.* 1. *Conch. tab.* 75. *fig. F.*

PLATE XLII.

Venus exoleta, antiquated. *Penn. Br. Zool. No.* 49. *tab.* 54. 55.— 49 & 49 *A.*

Pectunculus planus, crassus, striis capillaceis dense striatus. CAPILLACEUS, HAIR-STREAKED. *Da Costa. Brit. Conch. p.* 187. *sp.* 24.—*Tab.* 12. *fig.* 5. 5.

Found in plenty on several of the British shores, as Cornwall, Dorsetshire, Devonshire, and Yorkshire; also in the isle of Guernsey, and those of the Orkneys.

FIG. II.

VENUS SINUOSA.

A. INDENTED VENUS SHELL.

SPECIFIC CHARACTER.

Thin, convex, a deep obtuse *finus*, or bending on the front. *Penn. Br. Zool. p.* 95. *sp.* 51.

Figured and described by Pennant, from a specimen in the Portland cabinet, that was found at Weymouth. The shell we have figured is in the collection of the Rev. T. Rackett.

43

PLATE XLIII.

CYPRÆA PEDICULUS.

SEA LOUSE COWRY, or NUN.

GENERIC CHARACTER.

Sub-oval, blunt at the ends. Aperture, length of the shell, longitudinal, linear, toothed.

SPECIFIC CHARACTER

AND

SYNONYMS.

Convex, margined, and furrowed transversely across the back.

CYPRÆA PEDICULUS testa marginata transversim sulcata. *Linn Syst. Nat. p.* 1180. *No.* 364.

Concha veneris exigua, alba, striata, Nuns. *List. H. Conch. tab.* 707. *fig.* 57.

Concha veneris exigua purpascens, striis minimis transversis, tribus maculis fuscis dorso inspersa. The purple spotted nuns, *alias cowrie*, &c.—Concha veneris minima nullis maculis insignita. The smallest nuns without spots. *Borlase Cornw. p.* 277. *tab.* 28. *fig.* 12. 13.

Pou de Mer—*Argenv. Conch.* 1. *p.* 310. *tab.* 21. *fig. L. II. p.* 270. *tab.* 18. *fig. L.*

Porcellana vulgaris, parva, globosa, striata, candida, dorso sinuato. *Gualt.* 1. *Conch. tab.* 14. *fig. P.* & *tab.* 15. *fig. R.*

PLATE XLIII.

Cypræa pediculus, common. *Penn. Brit. Zool. No.* 82. *tab.* 70. *fig.* 82.

Cypræa exigua transversim striata, maculæ fuscæ dorso inspersa. Pediculus, seu monacha. *Da Costa Br. Conch. p.* 33. *sp.* 18. *tab.* 2. *fig.* 6. 6.

This shell is very common on our shores, and a variety of the same species is also abundant in the West Indies. The English shells of this kind are of various tints, generally whitish, and with or without spots; the exotic kind is distinguished by a furrow on the back.

44

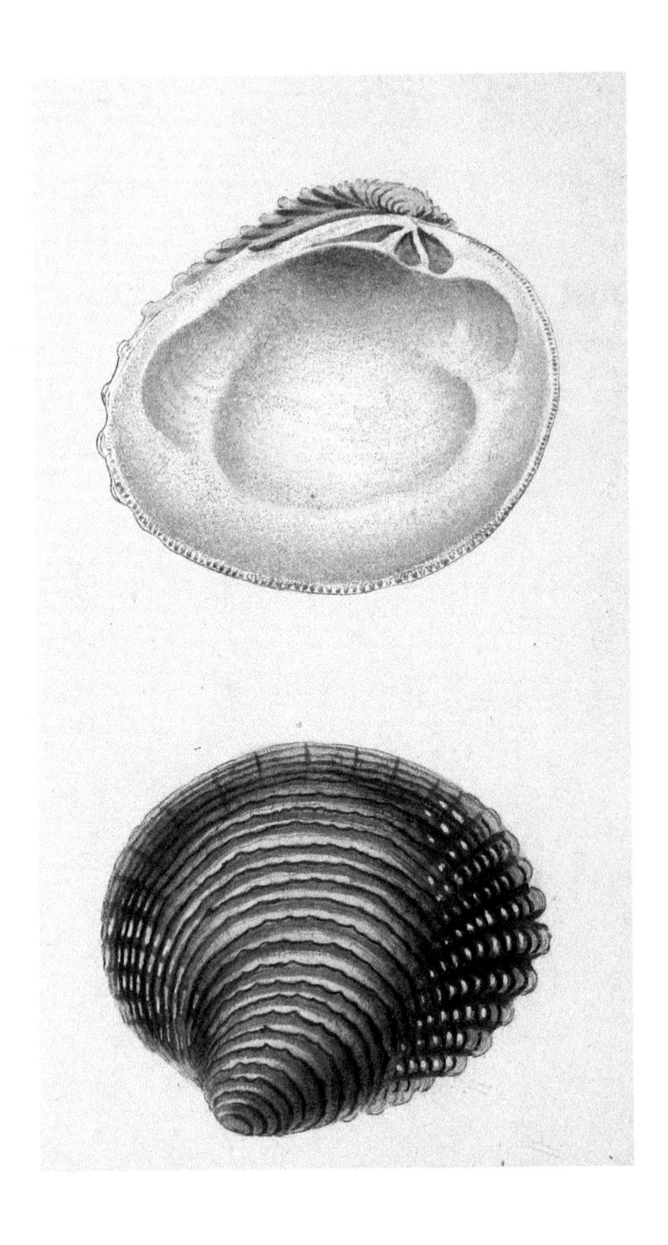

PLATE XLIV.

VENUS VERRUCOSA.

WARTED VENUS SHELL.

GENERIC CHARACTER.

Bivalve. Hinge furnished with three teeth; two near each other, the third divergent from the beaks.

SPECIFIC CHARACTER AND SYNONYMS.

Somewhat heart shaped. Deeply decussated on the sides, with transverse and oblique furrows, which form membraneous protuberances or warts. Margins finely crenated.

VENUS VERRUCOSA: testa subcordata: sulcis membranaceis striatis reflexis, antice imprimis, verrucosis, margine crenulato. *Linn. Syst. Nat. p.* 1130. *No.* 116.

Pectunculus omnium crassissimus, fasciis ex latere bullatis donatus, *List. H. Conch. tab.* 284. *fig.* 122.

Concha marina valvis æqualibus æqualitera, notabiliter umbonata et oblique incurvata, subrotunda, vulgaris, striis circularibus profundis, elatis, bullatis exasperata, et circumdata, crassa, subalbida. *Gualt.* 1. *Conch, tab.* 75. *fig. H.*

Concha cinerea densa, margine dentato, striis rugosis et é lateribus undose tuberculosis. The wrinkled, notched, and high-beaked concha, or cockle. *Borlase Cornw. p.* 278. *tab.* 28. *fig.* 32.

PLATE XLIV.

Clonisse de la Mediterranéé *d'Avila. Cab. p.* 333. *No.* 762.

Venus Erycina, Sicilian. *Penn. Brit. Zool. No.* 48. *tab.* 54. *fig.* 48.

Cornwall heart cockle, with rugged girdles. *Petiv. Gazoph. tab.* 93. *fig.* 17.

Pectunculus crassissimus strigatus, strigis ex latere bullatis, strigatus, ridged. *Da Costa Br. Conch. p.* 185. *sp.* 3. *Tab.* 12. *fig.* 1. 1.

Da Costa says, " this species is rare in our seas. The shores of Cornwall afford them, and they have been got in Devonshire and Dortsetshire." They have also been found on the eastern coast of Sussex, but not frequently.

45

PLATE XLV.

OSTREA STRIATA.

STRIATED OYSTER.

GENERIC CHARACTER.

Animal a Tethys. Shell bivalve, unequal. Hinge without a tooth, having a small oval cavity.

SPECIFIC CHARACTER AND SYNONYMS.

Less than the common Oyster. Outside wrought with thread-like longitudinal ridges. Inside green.

OSTREUM STRIATUM: mediæ magnitudinis veluti striatum intus virescente. *Da Costa. pl.* 11. *fig.* 4. 4. *p.* 162. *sp.* 9.

Ostreum parvum veluti striatum, testa intus virescente, cardine utrinque canaliculato. *List. H. An. Angl. p.* 181. *tit.* 27. *tab.* 4. *fig.* 27.

Ostrea fere circinata, subviridis, leviter striata. *List. H. Conch. tab.* 202. 203. *fig.* 36. 37.

An Ostreum vulgare, striatum, striis rotundis, crassioribus, interruptis radiatum, squamosum ex fusco viridescens. *Gualt.* 1. *Conch. tab.* 102. *fig. B?*

PLATE XLV.

" This Oyster," says Da Costa, " hitherto ***only proposed*** and described by Dr. *Lister*, is a very different species from the *common Oyster*, but has been always overlooked as the same kind." Dr. Lister observes that it is found in plenty at the mouth of the river *Tees*, in Yorkshire, and says he first eat of it at *Bourdeaux*, in France, where it is greatly esteemed and called Rock Oyster, being found among the rocks.

The figure of this shell in the plate of Da Costa above quoted, is so very indifferent and devoid of true character, that were we not in possession of the specimen he represents, it would be difficult to ascertain it. In the general description he says the outside is a little uneven, *but not rugged* nor of *a leaved or flakey structure* as the common Oyster: he adds that the ridges are longitudinal, about the thickness of a thread, very numerous, irregular, and run one into another; but towards the bottom always furcate or divide. This description is accurate but does not accord with the figure, in which the longitudinal ridges appear of a flakey structure or like laminæ, and not numerous, irregular threadlike striæ as in the shell, We have selected several characteristic specimens of this species in the annexed plate.

This shell is thick, strong, and nearly opake: it is usually about an inch in diameter; the valves unequal, the under one being very concave, the upper one flattish. Within, it is of a livid green and rather glossy, the hinge broad, deep, somewhat triangular and striated transversely. In many shells there is a remarkable white mark exactly resembling a thick spot of white oil paint, placed a little below the hinge, this spot always appears in radiated wrinkles from the centre, and is formed by the muscle of the shell.

PLATE XLV.

It is found on many of our shores, as Kent, Sussex, Dorsetshire, &c. in abundance, and of various colours; some are very fine like japan lacquer, and others of a violet, green, pink, yellow or pearly tint when much worn. It is remarkable, however, that the upper valves are so scarce, that hundreds of the lower valves are found to one of them.

46

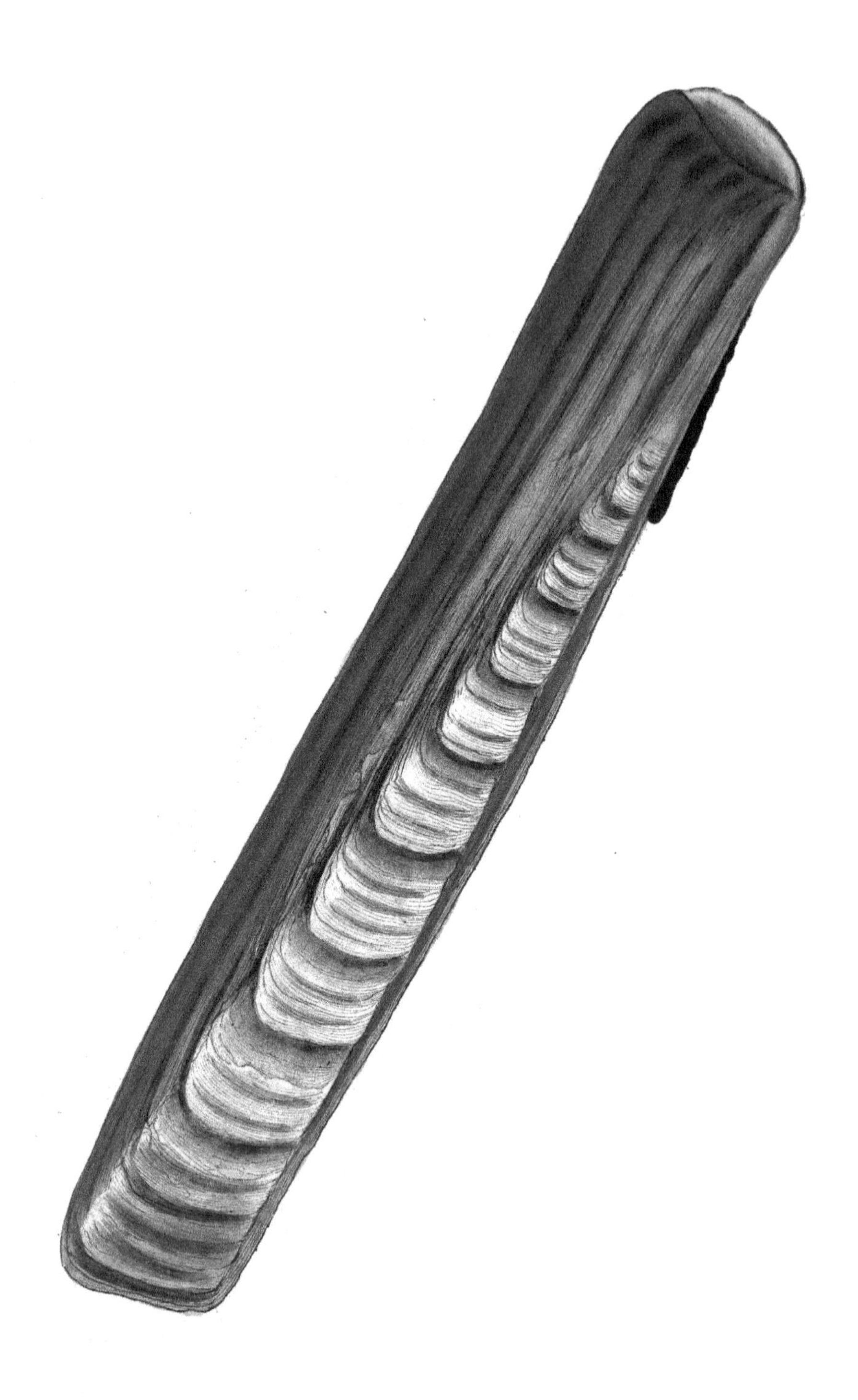

PLATE XLVI.

SOLEN SILIQUA.

LARGE or *POD SOLEN.*

GENERIC CHARACTER.

Bivalve, with equal valves, oblong, open at both ends. At the hinge a subulated tooth turned back, often double; not inserted in the opposite shell. Animal an Ascidia.

SPECIFIC CHARACTER

AND

SYNONYMS.

Shell strait, equally broad, and compressed. The hinge beset with two teeth in each valve.

SOLEN SILIQUA: testa lineari recta cardine altero bidentato. *Linn. Syst. Nat. p.* 1113. *No.* 34.—*Fn. Sv.* 2. *No.* 2131.

Solen, lævis, albidus, candidus, ex fusco & subroseo colore variegatus et fasciatus. *Gualt.* 1. *Conch. tab.* 95. *fig. C.*

Concha fusca, longissima, angustissimaque, musculo ad cardinem nigro, quibusdam solen dicta. *List. H. An. Angl. p.* 192. *tit.* 37. *tab.* 5. *fig.* 37.—*App. H. An. Angl. p.* 19.—*App. H. An. Angl. Goedart, p.* 33.

PLATE XLVI.

Solen major, subfuscus, rectus. *H. Conch. tab.* 409. *fig.* 255.

Solen unguis; the ſheath, razor, or spoutfish. *Grew Mus. p.* 143.
—*Merret Pin. p.* 193.

Solen sive concha tenuis longissimaque ab utraque parte naturaliter hians; the spout fish. *Wallace Orkneys, p.* 45.

Solen noſtras vulgaris. *Muſ. Petiv. p.* 87. *No.* 844.

Solen major subfuscus rectus. SILIQUA. *Da Costa. tab.* 17. *fig.* 5. *p.* 235. *sp.* 59.

This shell is found in abundance on many of the English shores, especially the northern and western coasts, and those of Scotland and Ireland.

The antients esteemed this fish a delicious food, and Dr. Lister informs us he thought it nearly as rich and palatable as the Lobster. In England and Scotland it is at present mostly used for baits, and not for the table; but in Ireland is much eaten in Lent.—It is in season in spring.

From the hinge to the opposite margin the length is about half an inch, and its breadth from five to seven inches; but some shells are found much larger. The outside is covered with a thin transparent yellow-brown cuticle or epidermis, like glue, which peels off soon after the fish is dead or exposed to the shores. Under this epidermis the shell is smooth, very glossy, and marked with many concentric transverse wrinkles from the middle to one extreme, the other half is striated lengthways. Inside white and glossy.

PLATE XLVI.

DA COSTA OBS.—Mr. Wallis, in his History of Northumberland, p. 396. No. 9. notes a sort of this shell he calls the *Orange and White Solen*, found in *Budle Sands* with this *common sort*, and in *all respects like it, except in colour*, which is deep orange and white in transverse fillets, in alternate variegations. *Quere, if a distinct species, or only a variety?*

D 2

47

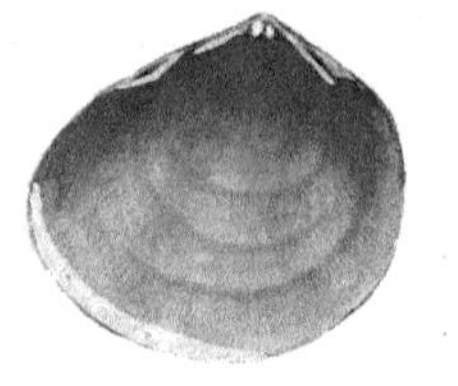

PLATE XLVII.

TELLINA CARNARIA. *Linn.*

FLESH-COLOURED TELLEN.

GENERIC CHARACTER.

The hinge usually furnished with three teeth; shell generally sloping on one side.

SPECIFIC CHARACTER.

Somewhat orbicular; valves shallow. White tinged with rose colour, and marked externally with numerous parallel striæ, disposed in an oblique, reflexed and transverse direction.

TELLINA CARNARIA: testa suborbiculata lævi utrinque incarnata oblique striata: striis hinc reflexis. *Linn. Syst. Nat. p.* 1119. *No.* 66.

Concha parva subrotunda, ex parte interna rubens. *List. H. An. Angl. p.* 175. *tit.* 25. *tab.* 4. *fig.* 5.

Tellina æquilatera lævis, tenuis subrubra. *Gualt.* 1. *Conch. tab.* 77. *fig.* 1.

Cardium parvum subrotundum oblique striatum colore carneoso. Carneosum. *Da Costa. p.* 181. *sp.* 20.

D 3

PLATE XLVII.

According to Dr. Lister this species is a shore shell, and found very frequently in the shallows of Lancashire, and near Filey in Yorkshire, &c. It is also found at Scarborough, Dorsetshire, Devonshire, and Cornwall.

It is an elegant shell, the outside being beautifully marked with numerous delicate striæ like strokes of engraving, and tinged with a fine rose or flesh colour. Some specimens are almost white, or white with transverse bands of deeper red, and the margins yellow. Within, the red colour is much more vivid than the outside.

Da Costa has placed this species in the *Cardium* genus, and indeed with much propriety; though Linnæus arranges it amongst the *Tellens*. It has a tendency on one side to flexure or slope like the Tellens, but the central and remote lateral teeth we think ſhould remove it to the former genus.

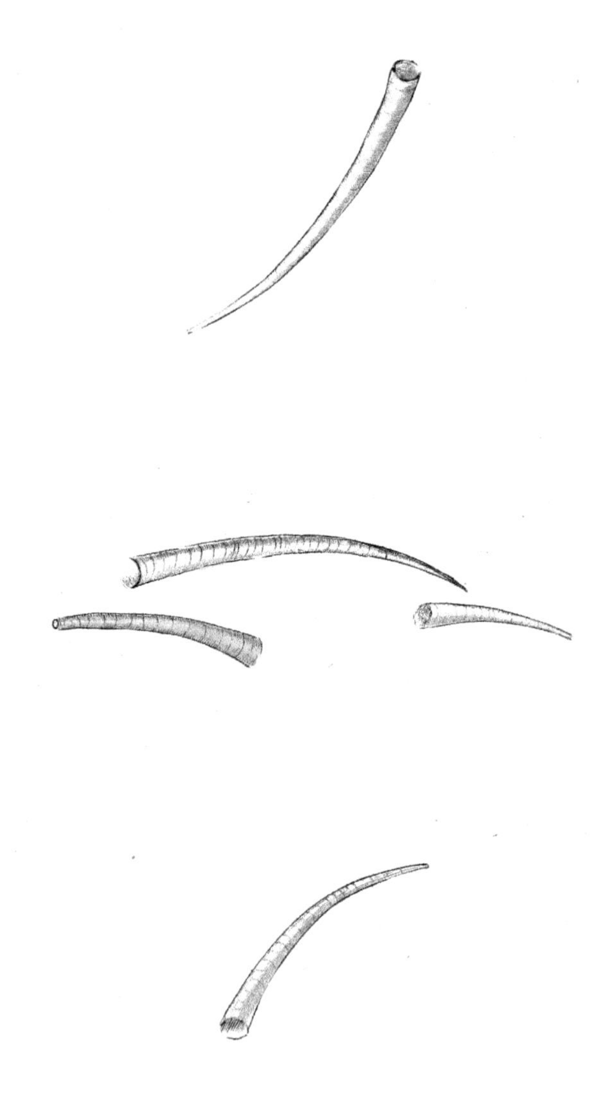

PLATE XLVIII.

DENTALIUM ENTALIS.

TOOTH SHELL.

GENERIC CHARACTER.

Animal a Terebella. Shell slender, tubiform.

SPECIFIC CHARACTER

AND

SYNONYMS.

Tubular, conical, somewhat curved, and open at both ends.

DENTALIUM ENTALIS: testa tereti subarcuata continua lævi. *Linn. Syst. Nat. p.* 3736. 332. *sp.* 3. *a Gmel.*

Dentale læve album, altera extremitate rufescens. *List. H. Conch. tab.* 547. *fig.* 2.

Tubulus marinus regulariter intortus arcuatim incurvatus, & versus unam extremitatem acuminatus, dentalis dictus, lævis, candidus. *Gualt.* 1. *Conch. tab.* 10. *fig. E.*

Dentale læve, curvum album. *Borlase Cornw. p.* 276. *tab.* 28. *fig.* 5.

PLATE XLVIII.

ANTALES *Argenville Conch.* 1. *p.* 246. *tab.* 7. *fig. K. II. p.* 196. *tab.* 3. *fig. K.*

Dentale læve albescens. Vulgare. *Da Costa. Br Conch. p.* 24. *tab.* 2. *fig.* 10.

This singular shell is found on many of the British shores, but rarely occurs perfect. It is abundant on our southern shores, as Hampshire, Devonshire, &c.

49

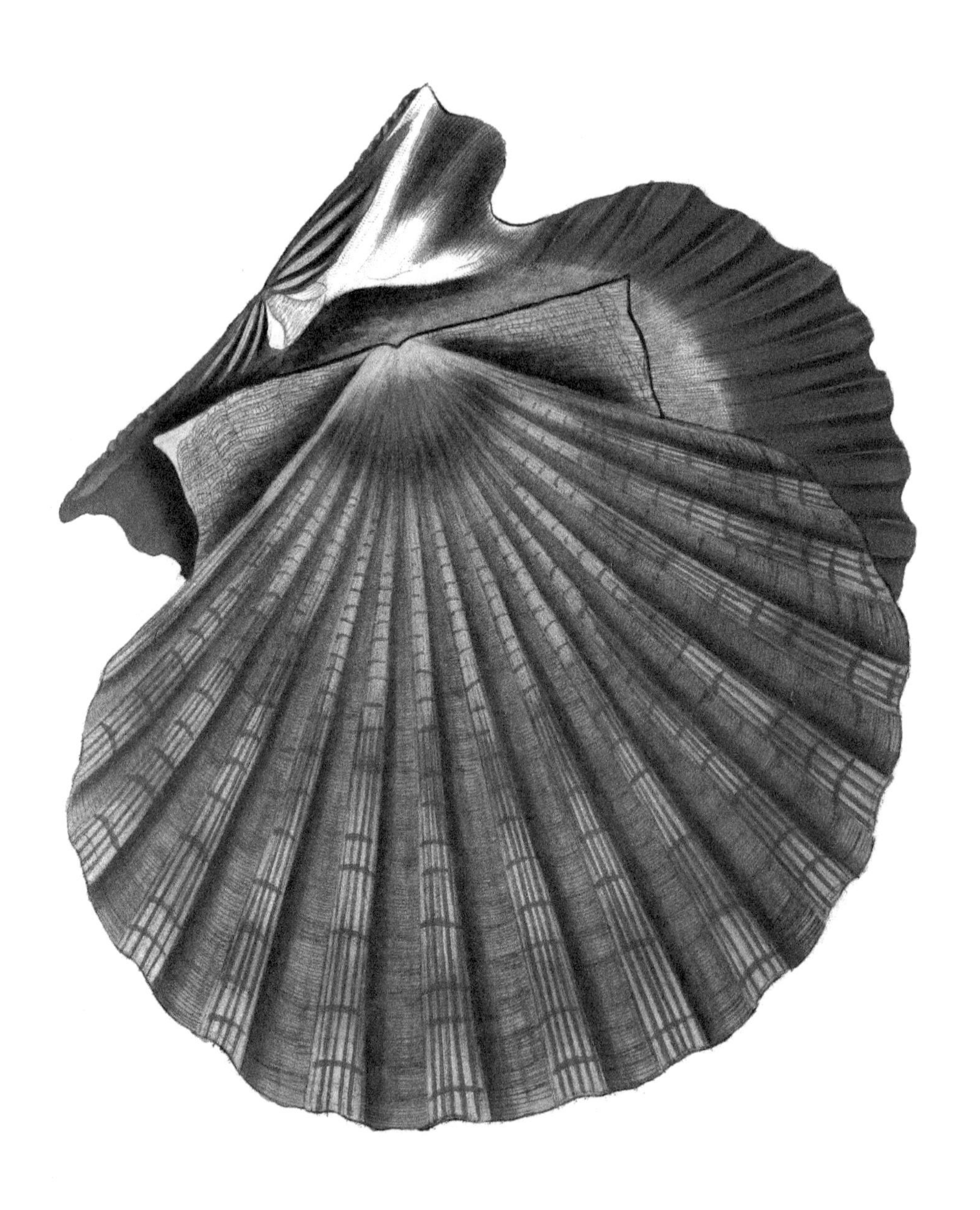

PLATE XLIX.

OSTREA MAXIMA.

GREAT SCALLOP.

GENERIC CHARACTER.

Animal a Tethys. Shell bivalve, unequal. Hinge without a tooth, having a ſmall oval cavity.

SPECIFIC CHARACTER

AND

SYNONYMS.

Upper valve flat, lower concave. About fourteen rounded longitudinal ribs, which are also deeply striated.

OSTREA MAXIMA: testa inæquivalvi radiis rotundatis longitudinaliter striatis. *Linn. Syst. Nat. p.* 1144 *No.* 185.

Concha testa aurita, striis quatuordecim. *Linn. F. Suec. I. p.* 383. *No.* 1343. *II.* 2148.

P. maximus, circiter quatuordecim striis, admodum craſſis & eminentibus et iisdem ipsis striatis inſignitus. A Scallop. *List. H. An. Angl. p.* 184. *tit.* 29. *tab.* 5. *fig.* 29.

Escallop, or Scallop. *Merret. Pin.* 193.

Scallop or Clam-shell. *Wallace Orkn. p.* 43. &c.

PLATE XLIX.

Frill or Scallop. *Hutchins Dorset. p.* 77.

Pecten Maximus. Great. *Penn. Br. Zool. No.* 61. *tab.* 59. *fig.* 61.

Pecten. Maximus inæquivalvis, costis circiter quatuordecim rotundatis, & admodum crassis. Vulgaris, the common scallop. *Da Costa Br. Conch. p.* 140. *tab.* 9. *fig.* 3. 3.

The large Escallop is found on most of the coasts of Great Britain and Ireland, particularly on those of Portland and Purbeck in Dorsetshire.—The fish is eaten and much esteemed.

It is said by modern, as well as antient authors, that Escallops will move so strongly as to leap out of the catcher wherein they are taken: their way of leaping or raising themselves up, is by forcing the under valve against whatever they lie upon.

50

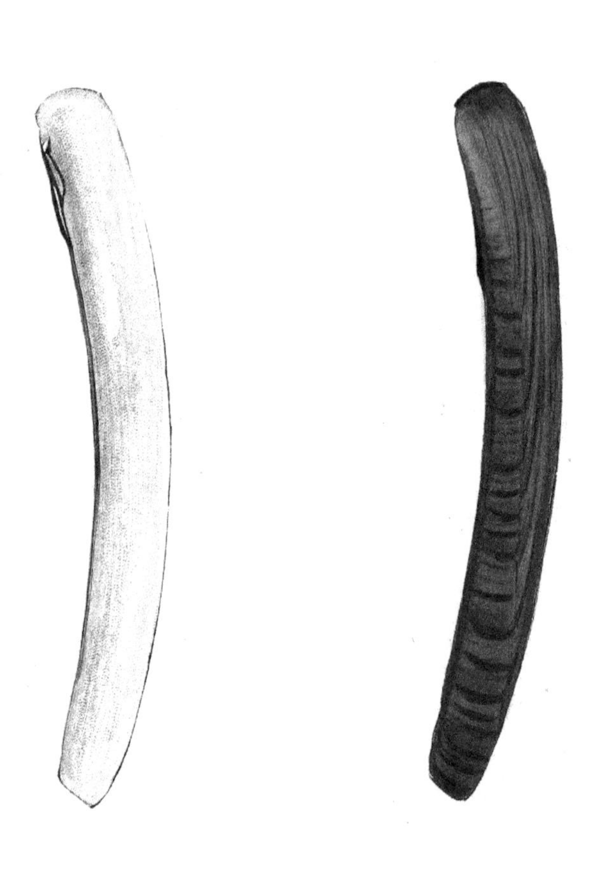

PLATE L.

SOLEN ENSIS.

SCYMETAR.

GENERIC CHARACTER.

Bivalve, with equal valves, oblong, open at both ends. At the hinge a subulated tooth turned back, often double; not inserted in the opposite shell.—Animal an Ascidia.

SPECIFIC CHARACTER

AND

SYNONYMS.

Shell bowed like a Scymetar, equally broad and compressed. The hinge beset with two teeth in each valve.

Solen ensis: tefta lineari fubarcuata, cardine altero bidentato. *Linn. Syst. Nat. p.* 1114. *No.* 35.

Solen alter curvus minor. *List. App. H. An. Angl. p.* 20.—*App. in Goed. p.* 36. *tab.* 2. *fig.* 9.—*Solen curvus. Hift. Conch. tab.* 411. *fig.* 257.—Solen ensis, Scymetar. *Penn. Br. Zool. No.* 22. *tab.* 45. *fig.* 22.

Solen subarcuatus. Ensis. *Da Costa. Br. Conch. p.* 237. *sp.* 60.

PLATE L.

This is a local and rare species; it has been found at Weymouth on the Dorsetshire coast, and according to Dr. Lister in plenty in the *æstuary* of the Severn, on the side of Wales.

It differs from the *Solen siliqua* in several respects; it is smaller, and not strait, but curved or bowed like a Scymetar.

51

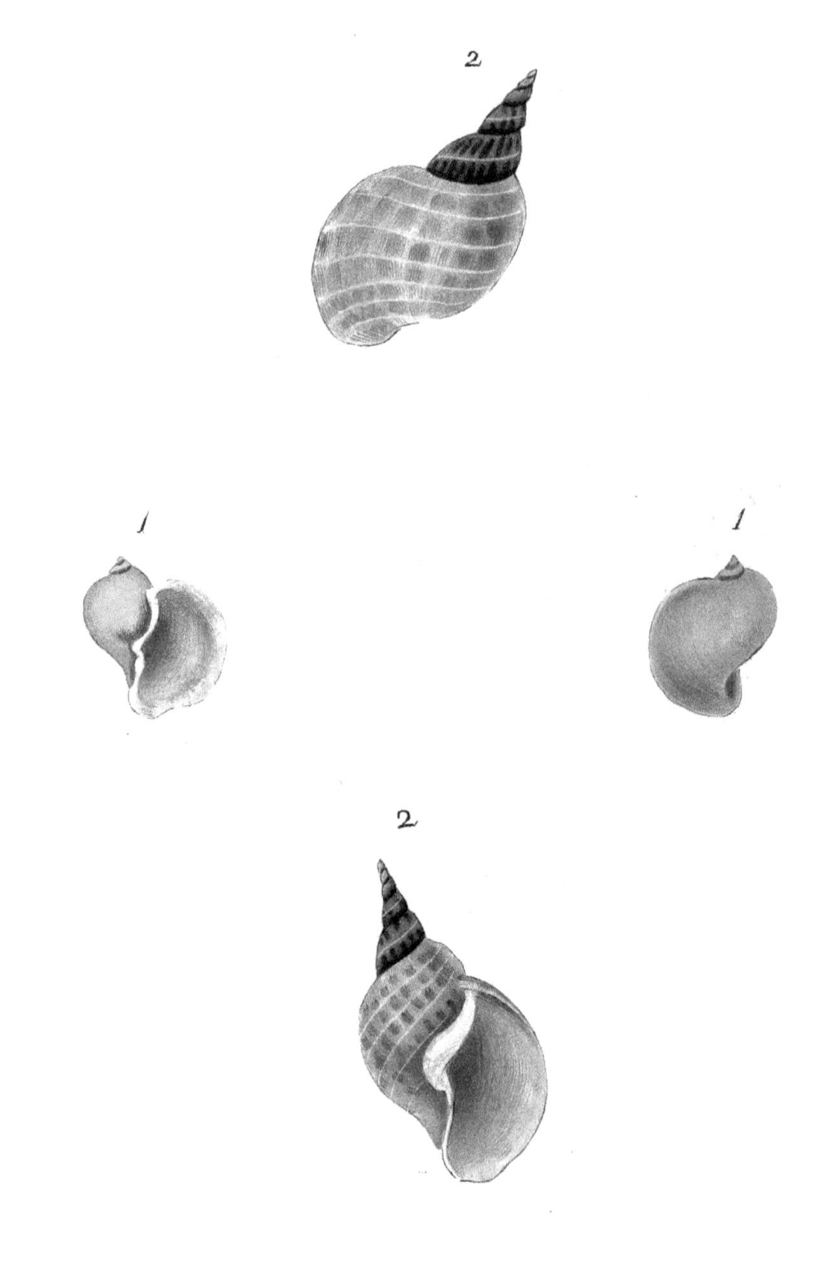

PLATE LI.

FIG. II.

HELIX AURICULARIA.

EAR, or WIDE MOUTH RIVER SNAIL.

GENERIC CHARACTER.

Aperture of the mouth lunated.

**** *Ovated, imperforated.*

SPECIFIC CHARACTER

AND

SYNONYMS.

Without umbilicus: a narrow depression on the edge of the pillar lip. Very ventricose, spire short and acute. Aperture very wide.

HELIX AURICULARIA: testa imperforata ovata obtusa, spira acuta brevissima, apertura ampliata. *Linn. Syst. Nat.* *p.* 1250. *No.* 708.

Cochlea testa diaphana anfractibus quatuor, mucrone acuto brevissimo, apertura acutissima. *Linn. F. Suec. I.* *p.* 376. *No.* 1315. *II. No.* 2192.

Buccinum pellucidum subflavum, quatuor spirarum, mucrone acutissimo, testæ apertura omnium maxima. *List. H. An. Angl. p.* 139. *tit.* 23. *tab.* 2. *fig.* 23.

PLATE LI.

Buccinum subflavum pellucidum, quatuor orbium, ore amplissimo, mucrone acuto. *List. H. Conch. tab.* 123. *fig.* 22.

Buccinum fluviatile pellucidum, subflavum, quatuor spirarum, mucrone acuto, testæ apertura patentissima. *List. Exerc. Anat.* 2. *p.* 54.

Turbo with four wreaths, a remarkable large mouth, and a short acute apex. *Wallis Northumb. p.* 370.

Helix auricularia. *Penn. Br. Zool. No.* 138. *tab.* 86. *fig.* 138.

Turbo subflavus pellucidus quatuor spirarum ore patulo. Patulus. *Da Costa sp.* 53. *p.* 95. *tab.* 6. *fig.* 17.

Found in plenty in rivers, ponds, &c.

FIG. II.

HELIX STAGNALIS,

LAKE SNAIL, or FRESH WATER TURBO.

SPECIFIC CHARACTER

AND

SYNONYMS.

Without umbilicus. Oblong; spire tapering. Several prominent longitudinal wrinkles which somewhat angulates the shell. Aperture oblong oval.

PLATE LI.

Helix Stagnalis: testa imperforata ovata-subulata, subangulata, apertura ovata. *Linn. Syst. Nat. p.* 1249. *No.* 703.

Cochlea testa producta acuminata, opaca, anfractibus senis subangulatis, apertura ovata. *Linn. F. Suec. I. p.* 374. *No.* 1310. *II. No.* 2188.

Buccinum longum sex spirarum, omnium & maximum & productius, subflavum, pellucidum, in tenue acumen ex amplissima basi mucronatum. Turbo lævis in stagnis degens. *Aldror. de Testaceis, I.* 3. *p.* 359. *No.* 3.

Buccinum subflavam pellucidum, sex orbium, clavicula admodum tenui, productiore. *List. H. Conch. tab.* 123. *fig.* 21.

Buccinum minus fuscum, sex spirarum, ore angustiore. *List. H. An Angl. p.* 139. *tit.* 22. *tab.* 2. *fig.* 22.

Helix Stagnalis. *Lake. Penn. Br. Zool. No.* 136. *tab.* 86. *fig.* 136.

Fresh water turbo with six wreaths. *Wallis Northumb. p.* 369.

Turbo longus et gracilis in tenue acumen mucronatus, imperforatus & pellucidus Stagnalis. *Da Costa Br. Conch. p.* 93. *sp.* 52. *tab.* 5. *fig.* 11.

The largest and most produced of the British river snails, and is found in plenty in all our rivers, lakes, ponds, and other waters.

Lifter and Petiver have made two species of this shell maximum and minus; they appear however to be merely different stages of its growth.

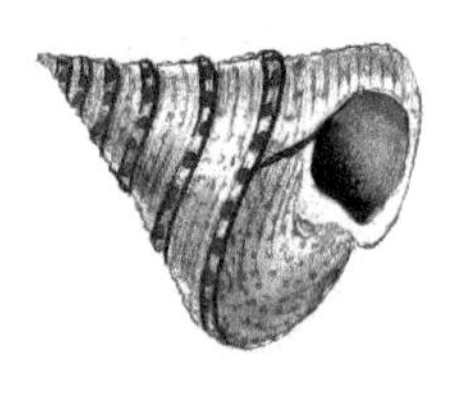

PLATE LII.

TROCHUS ZIZYPHINUS.

LIVID TOP SHELL.

GENERIC CHARACTER.

Animal a slug. Shell conic. Aperture nearly triangular.

SPECIFIC CHARACTER

AND

SYNONYMS.

Shell conic, livid, without umbilicus: spirally striated, with the upper edge of each wreath margined.

TROCHUS ZIZYPHINUS: testa imperforata conico livida lævi, anfractibus marginatis. *Linn. Syst. Nat. p.* 1231. *No.* 599.—*Faun. Suec. II. No.* 2168.

Trochus albidus maculis rubentibus distinctus, sex minimum spirarum. *List. H. An. Angl. p.* 166. *tit.* 14. *tab.* 3. *fig.* 14.

Trochus pyramidalis variegatus, limbo angusto in summo quoque orbe circumdatus. *List. H. Conch. tab.* 616. *fig.* 1.

Culs de Campe de moyenne grandeur, lisses, marbrès de rouse et de violet, à orbes separès par un cordon. *D'Avila, cab. p.* 127. *No.* 155.

PLATE LII.

Trochus Ziziphinus, livid. *Penn. Br. Zool. No.* 103. *tab.* 80. *fig.* 103.

Trochus pyramidalis imperforatus, lividus, rubro variegatus, limbo in summo quoque orbe circumdatus. Zizyphinus. *Da Costa Br. Conch. tab.* 3. *fig.* 2. 2. *p.* 37.

This is one of the moſt elegant of the testaceous tribe found on our coasts; the colour in general is fleſh colour or pale red, elegantly variegated with deeper red or brown in streaks, waves, and chequers; when the exterior coat is worn, the shell is of a fine pearly hue.

It is not an uncommon species on the English shores, and is also found in the Orkneys and the western isles of Scotland.

53

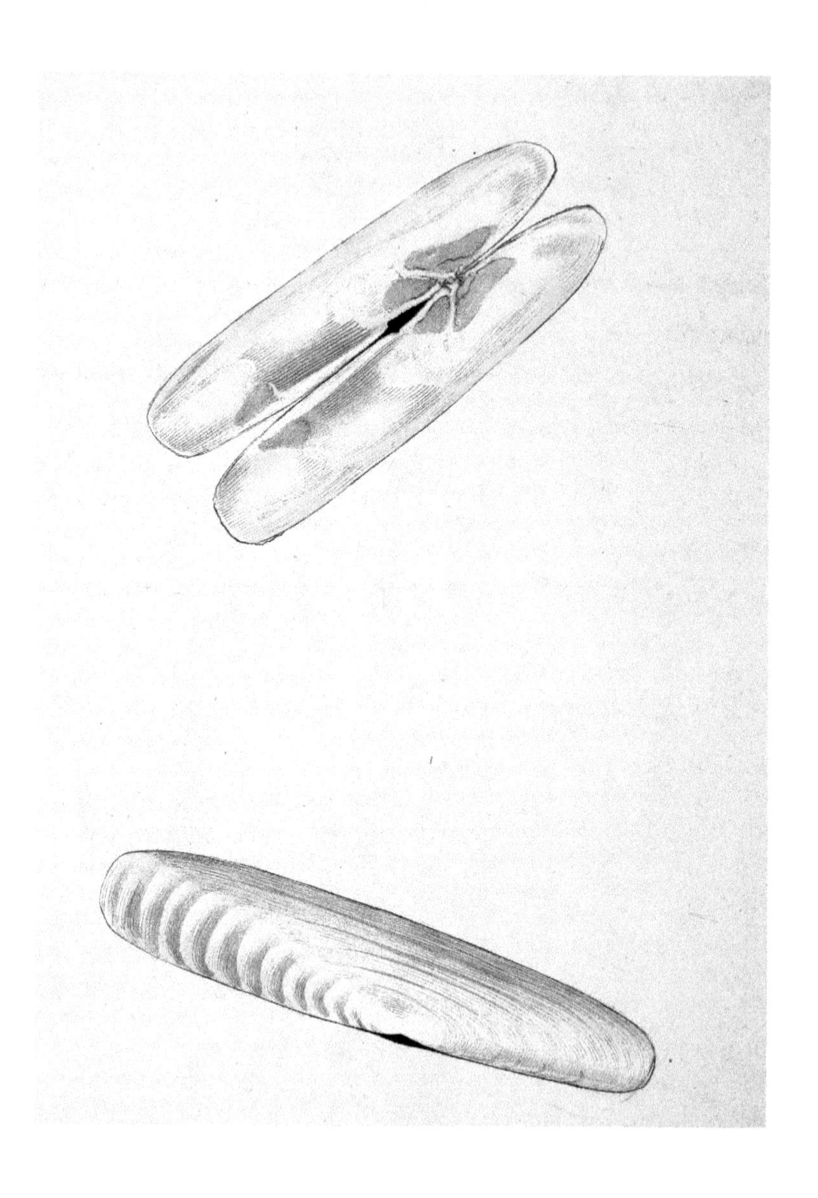

PLATE LIII.

SOLEN LEGUMEN.

PEASECOD.

GENERIC CHARACTER.

Bivalve, with equal valves, oblong, open at both ends. At the hinge a subulated tooth turned back, often double; not inserted in the opposite shell. Animal an afcidia.

SPECIFIC CHARACTER

AND

SYNONYMS.

Strait, oblong, rounded at both ends: smooth, and somewhat pellucid.

SOLEN LEGUMEN: curtus subpellucidus, ad chamas quodammodo accedens. Peasecod. *Da Costa. Br. Conch. p.* 238. *sp.* 61.

Solen Legumen, Suboval. *Penn. Br. Zool. No.* 24. *tab.* 46. *fig.* 24.

Concha soleniformis, lævis aut levissime striata, fragilis, pellucida, testa tenuissima cornea, subalbida, aliquando flavescens. *Gualt.* 1. *Conch. tab.* 91. *fig. A.*

Chama subfusca, angustissima, ad solenes quodammodo accedens. *List. II. Conch. tab.* 420. *fig.* 264.

E 2

PLATE LIII.

Both Pennant and Da Costa note this as a very rare British species. The first says it is found at Red Wharf, Anglesea, in North Wales; the latter received it from Christchurch, in Hampshire.

We have found it on the shores of Glamorganshire, and also in abundance in the sandy bay of Caermarthen this summer.

54

PLATE LIV.

CARDIUM LÆVIGATUM.

LARGE HIGH-BEAKED COCKLE.

GENERIC CHARACTER.

Two teeth near the beak, and another remote one on each side of the shell.

SPECIFIC CHARACTER

AND

SYNONYMS.

Shell somewhat oval, slightly striated longitudinally.

CARDIUM LÆVIGATUM: testa obovata: striis obsoletis longitudinalibus. *Gmel. Linn. Syst. Nat. p.* 3251. *sp.* 18.

Pectunculus maximus, at minus concavus; plurimis minutioribus & parum eminentibus striis donatus, rostro acuto, minusque incurvato. *List. H. An. Angl. p.* 187. *tit.* 32. *tab.* 5. *fig.* 32.

Pectunculus subfuscis striis leviter tantum incisis. *List H. Conch. tab.* 332. *fig.* 169.

Pectunculus major striis angustis. *Petiv. Gaz. tab.* 93. *fig.* 10.

Large high-beaked Cockle. *Wallis Northumb. p.* 395.

Cardium Lævigatum. Smooth. *Penn. Br. Zool. No.* 40. *tab.* 51. *fig.* 40.

Cardium obovatum striis obsoletis longitudinalibus. LÆVIGATUM. *Da Costa. Br. Conch. p.* 178. *sp.* 18.

E 3

PLATE LIV.

We have observed, that this species is in general discoloured, and deeply tinged with brown or yellow ; when fine it is whitish, sleek, and covered with an epidermis.

It is found on most of our coasts, yet it is by no means common. Da Costa notes it from Yorkshire, Northumberland, Dorsetshire, Cornwall, Carnarvonshire, and the Orkneys.

55

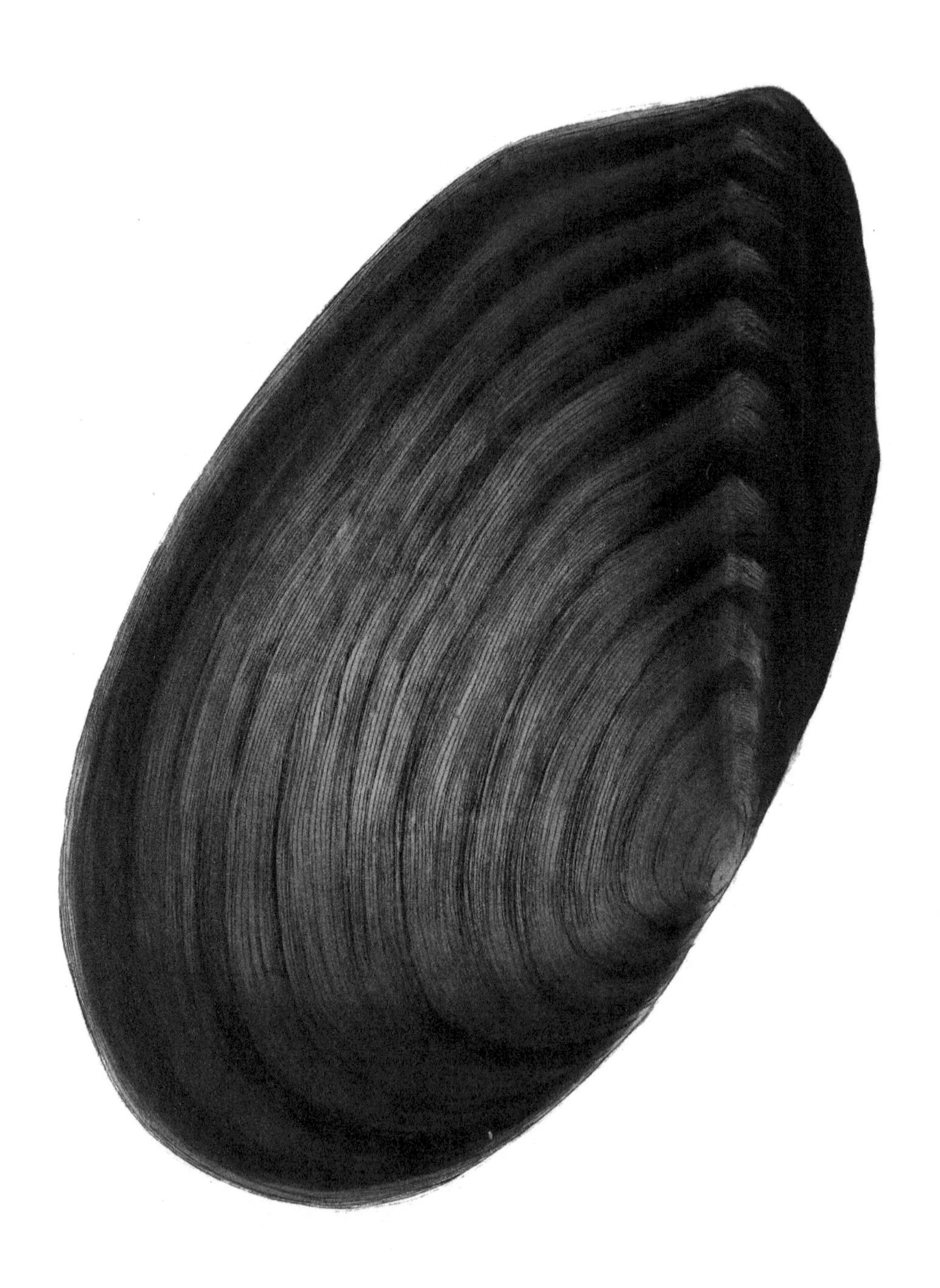

PLATE LV.

MYTILUS CYGNEUS.

GREAT HORSE, OR SWAN MUSCLE.

GENERIC CHARACTER.

The hinge toothless, and consists of a longitudinal furrow.

SPECIFIC CHARACTER

AND

SYNONYMS.

Somewhat oval, one end rounded, the other extended, narrow, and compressed; thin and brittle.

MYTILUS CYGNEUS: testa ovata anterius compressiuscula fragilissima, cardine laterali. *Gmel. Linn. Syst. Nat. p.* 3355. 15.

MYTILUS CYGNEUS: concha testa oblonga ovata longitudinaliter subrugosa, postice compresso-prominulo. *Linn. Fn. Suec.* 1. *p.* 380. *No.* 1332.

Musculus latus maximus, testa admodum tenui, ex fusco viridescens, palustris. *List. App. H. An. Angl. p.* 8. *tit.* 30. *tab.* 1. *fig.* 3. *and App. H. An. Angl. in Goedart. p.* 9. *tit.* 30. *tab.* 1. *fig.* 3.—Musculus latus maximus & tenuissimus é cœruleo viridescens, fere palustris. *H. Conch. tab.* 156. *fig.* 11.

E 4

PLATE LV.

Musculus fluviatilis maximus, profunde striatus latus; testa admodum tenui, ex fusco viridescens, interdum rufescens, intus argenteus. *Gualt.* 1. *Conch. tab.* 7. *fig. F.*

Mytilus Cygneus. Swan. *Penn. Brit. Zool. No.* 78. *tab.* 67. *No.* 78.

Mytilus fluviatilis maximus, admodum tenuis ex fusco viridescens. CYGNEUS. *Da Costa Br. Conch. p.* 214. *sp.* 46.

This species is not uncommon in our Rivers, Ponds, &c. but is less frequent than the Mytilus Anatinus, or Small Horse Muscle, which bears some resemblance to it. Mytilus Anatinus is rarely more than half the size of Mytilus Cygneus, is more compressed, and has the cartilage side extended in a straight line to an acute angle at one end.

The usual length of Mytilus Cygneus is about two or three inches, its breadth five or six inches. The valves deep, or concave. The outside is wrinkled transversely, and varies in colour according to the state of the Shell. The external covering, or epidermis, is thin, but strong, and of a greenish colour; under this the Shell is varied with bright brown, and when the coating is worn off, the whole Shell is pearly. The inside is often rugged with small pearls.

56

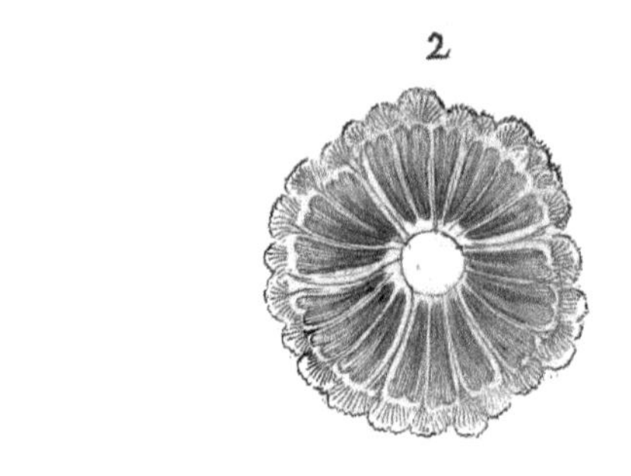

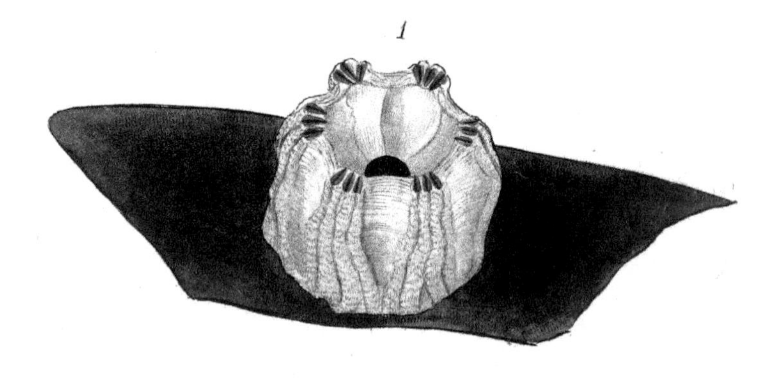

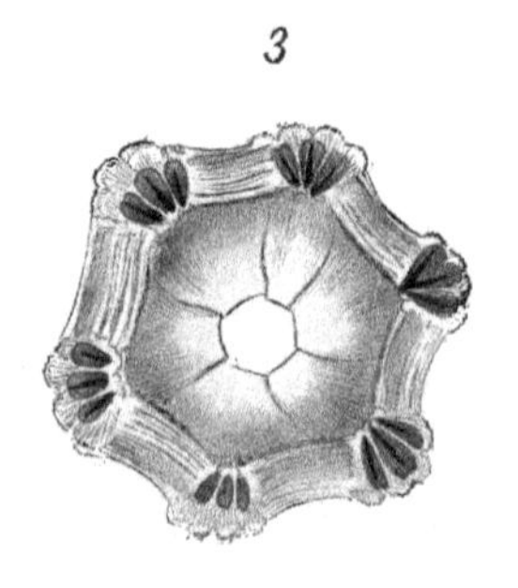

PLATE LVI.

LEPAS DIADEMA.

WHALE ACORN SHELL.

GENERIC CHARACTER.

Animal Triton. Shell of many unequal valves; affixed by a stem.

SPECIFIC CHARACTER

AND

SYNONYMS.

Shell subrotund, of six lobes, furrowed longitudinally.

LEPAS DIADEMA: testa subrotunda, sexlobata sulcata. *Gmel. Linn. Syst. Nat. p.* 3208. *sp.* 4.

Balanus balænæ cuidam Oceani Septentrionalis adhærens. *List. H. Conch. tab.* 445. *fig.* 288.

Pediculus ceti. *Phil. Trans. No.* 222. *p.* 323. *Epitome Trans. Soc. R. Angl. Vol.* 5. *p.* 381. *tab.* 17. *fig.* 2.

Pediculus ceti, vel Lepas nuda carnosa aurita. *Idem.* 1758. *Vol.* 50. *P.* 2. *tab.* 34. *fig.* 1. *and fig.* 7.

Martin. *West. Isles, p.* 162 *and* 166.

Quarta species echini plani. *Rumph. Mus. tab.* 14. *fig. H.*

Balanus balænaris. *Klein. Ostrac.* 176. *tab.* 12. *fig.* 98.

Pediculus ceti. *Boccon. recher. p.* 287.

Grand pou de Baleine. *D'Avil. Cab. p.* 404.

Balanus hemisphericus sexlobatus. Balænæ. *Da Costa Br. Conch. tab.* 17. *fig.* 2. 2. 2.

PLATE LVI.

This large and interesting species of Balani is found adhering to the Whale, whence it is called the Whale Acorn Shell. It is not uncommon in the sea round Scotland. The natives of some of the Western Islands distinguish one species of Whale from the rest, for its great size, and the *big limpets* growing on their backs *. The same species is common on the Whales in the Northern Seas about Newfoundland.

The Animal is figured by Ellis, in the Philosophical Transactions for 1758, and resembles a cluster of small hooded and eared serpents issuing from the central cavity, and little openings at the tops of the longitudinal ribs. The base by which it is affixed, when alive, is covered with a coriaceous skin.

---

* Martin. Fladda Chuan, near the Isle of Skie.

57

PLATE LVII.

VOLUTA TORNATILIS.

OVAL VOLUTE.

GENERIC CHARACTER.

Animal Limax. Shell spiral, aperture narrow, without a beak. Columella plicated.

SPECIFIC CHARACTER
AND
SYNONYMS.

Shell oval, pointed at each end, and striated spirally. Pillar lip turned in a fold.

VOLUTA TORNATILIS: testa coarctata ovata substriata: spira elevata acutiuscula, columella uniplicata. *Gmel. Linn. Syst. p.* 3437.

Buccinum parvum, rostro integro, tenuiter striatum, fasciatum, clavicula paulo productiore, unico dente ad columellam. *List. H. Conch. tab.* 835. *fig.* 58.

Auris Midæ fasciata. *Klein Ostrac. p.* 37. §. 96. *sp. I. No. I.*

Voluta tornatilis. Oval. *Penn. Br. Zool. No.* 86. *tab.* 71. *fig.* 86. Schroet. *n.* Litterat. 3. *t.* 2. *f.* 12. 13.

Ovalis. Turbo ovalis striatus rubicundus fasciis albis, columella uniplicata & unidentata. *Da Costa Br. Conch. p.* 101. *tab.* 8. *fig.* 2. 2. *sp.* 57.

PLATE LVII.

" This pretty species," says Da Costa, " I have received from Tinmouth and Exmouth, in Devonshire;" and Pennant notes it from Anglesea only.

58

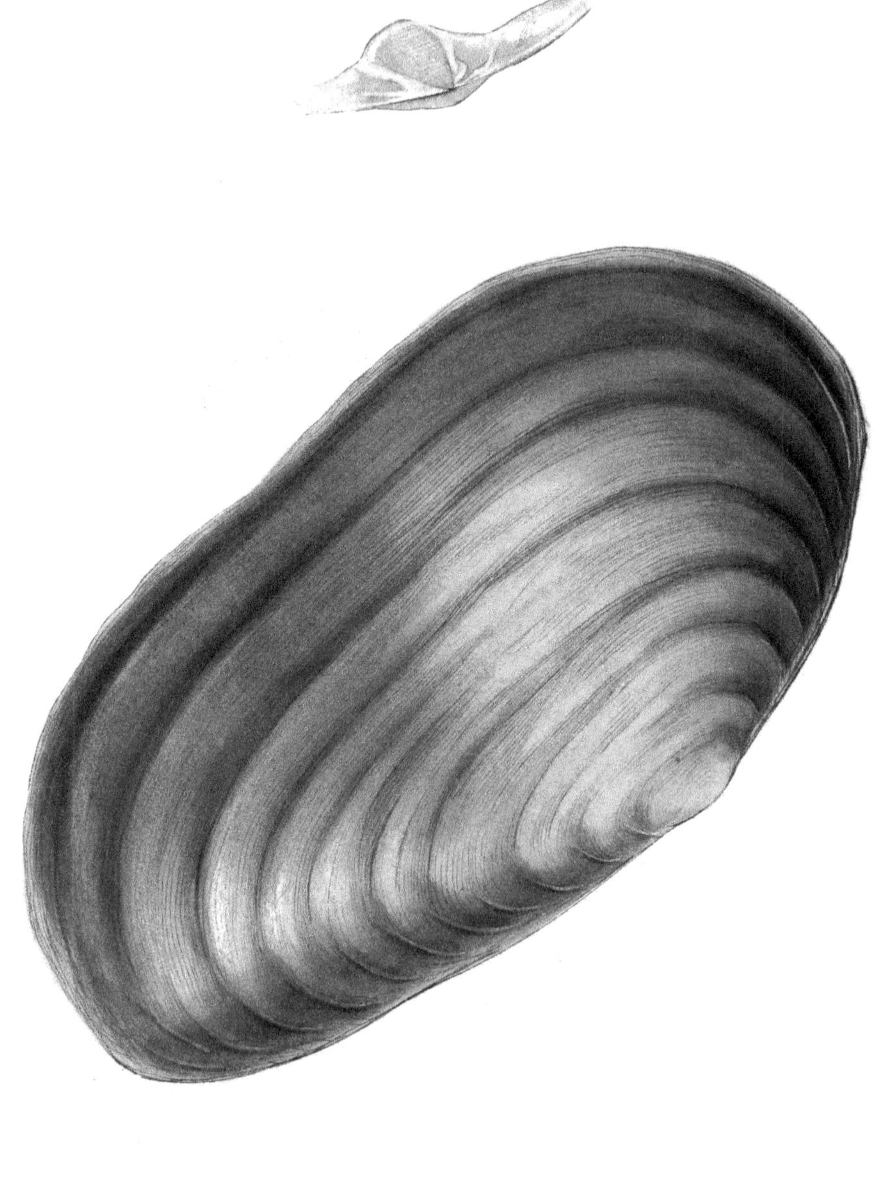

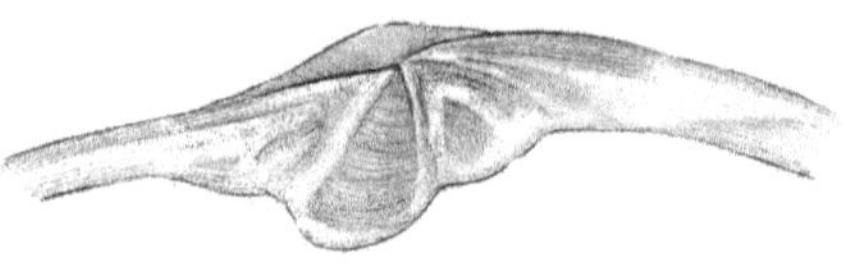

PLATE LVIII.

MACTRA LUTRARIA.

GENERIC CHARACTER.

Animal a Tethys. Bivalve, sides unequal. Middle tooth complicated, with a little groove on each side; lateral teeth remote.

SPECIFIC CHARACTER AND SYNONYMS.

Shell oblong oval, smooth; no lateral teeth. Hinge, a small and large triangular cavity in one valve; a similar cavity and an elevated triangular tooth in the opposite.

MACTRA LUTRARIA: testa ovali oblonga lævi, dentibus lateralibus nullis. *Gmel. Linn. Syst. p.* 3259. *sp.* 14.

Concha longa lataque in mediis cardinibus cavitate quadam pyriformi insignita.—An Chamæ glycymeris Rondeletii? *List. H. Angl. p.* 170. *tit.* 19. *tab.* 4. *fig.* 19.—Chama fusca lata planior. *Hist. Conch. tab.* 415. *fig.* 259.

The long and broad conch. *Wallis Northumb. p.* 396. *No.* 10. 11.

Mactra lutraria, large. *Penn. Br. Zool. No.* 44. *tab.* 52. *fig.* 44. *Chemnitz. Conch.* 6. *t.* 24. *f.* 240. 241.

Chama magna planior, crassa, albescens, Magna. *Da Costa Br. Conch. p.* 230. *sp.* 55. *tab.* 17. *fig.* 4.

PLATE LVIII.

The Mactra lutraria is so very similar in general appearance to the Mya Arenaria, that without attending to the foliated hinge of the latter, they may be confounded with each other. Both shells are scarce on the British coasts, except in certain situations. Da Costa says the Mactra lutraria is found in plenty at Scarborough, in Northumberland, Lancashire, &c. and on the shores of Scotland. Dr. Maton found them on the coast of Cornwall; and we met with them very fine, perfect, and beautifully coloured, on the sands near Laugharn, South Wales.

The general colour is yellowish, tinged with orange, and irregularly clouded with brown; and often glossed with a delicate white; the epidermis brown.

59

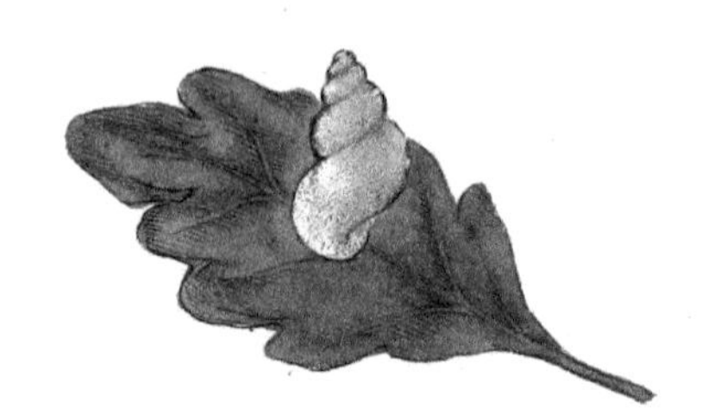

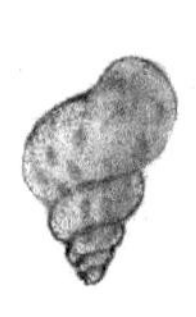

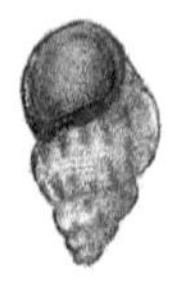

PLATE LIX.

TURBO STRIATUS.

STRIATED WREATH SHELL.

GENERIC CHARACTER.

Animal Limax. Univalve, spiral, or of a taper form. Aperture somewhat compressed, orbicular, entire.

SPECIFIC CHARACTER

AND

SYNONYMS.

Shell swelled, or ventricose, white, variegated with red, and finely striated spirally. Umbilicated.

TURBO STRIATUS: albescens rufo variegatus, eleganter striatus. *Da Costa. Br. Conch. p.* 86. *sp.* 47. *tab.* 5. *fig.* 9.

Turbo reflexus: testa umbilicata convexo-prominula: anfractibus teretibus substriatis, apertura reflexa. *Gmel. Linn. Syst. Nat. p.* 3605. 70?

Cochlea cinerea, interdum leviter rufescens, striata, operculo testaceo cochleato donata.—Cochlea terrestris turbinata et striata Columnæ de purpura. *c.* 9. *p.* 18. ubi etiam delineatur sub hoc titulo, Cochlea turbinata. *List. H. An. Angl. p.* 119. *tit.* 5. *tab.* 2. *fig.* 5.

Turbo lunaris tessellatus et striatus. *Klein Ostrac. p.* 55. §. 161. *spec.* 3.

PLATE LIX.

Argenv. Conch. *I. p.* 384. *tab.* 32. *fig.* 12. *II. p.* 339. *tab.* 28. *fig.* 12.

Turbo terrestris tenuissime, ipso ore circinato, cui etiam limbus latus, et striatus, albidus, *Gualt I. Conch. tab.* 4. *fig. B.*

Turbo tumidis. Tumid. *Penn. Br. Zool. No.* 110. *tab.* 82. *fig.* 110.

This species is particularly noticed by most conchologists. Dr. Lister says it is the most elegant of all our snails, and is found near Oglethorpe and Burwell woods in Lincolnshire, in Yorkshire, and in Kent. Petiver found it about Charlton, in Kent; also Morton, in Northamptonshire; Pennant, in the woods of Cambridgeshire; and Da Costa, in Surrey. It is no where common.

60

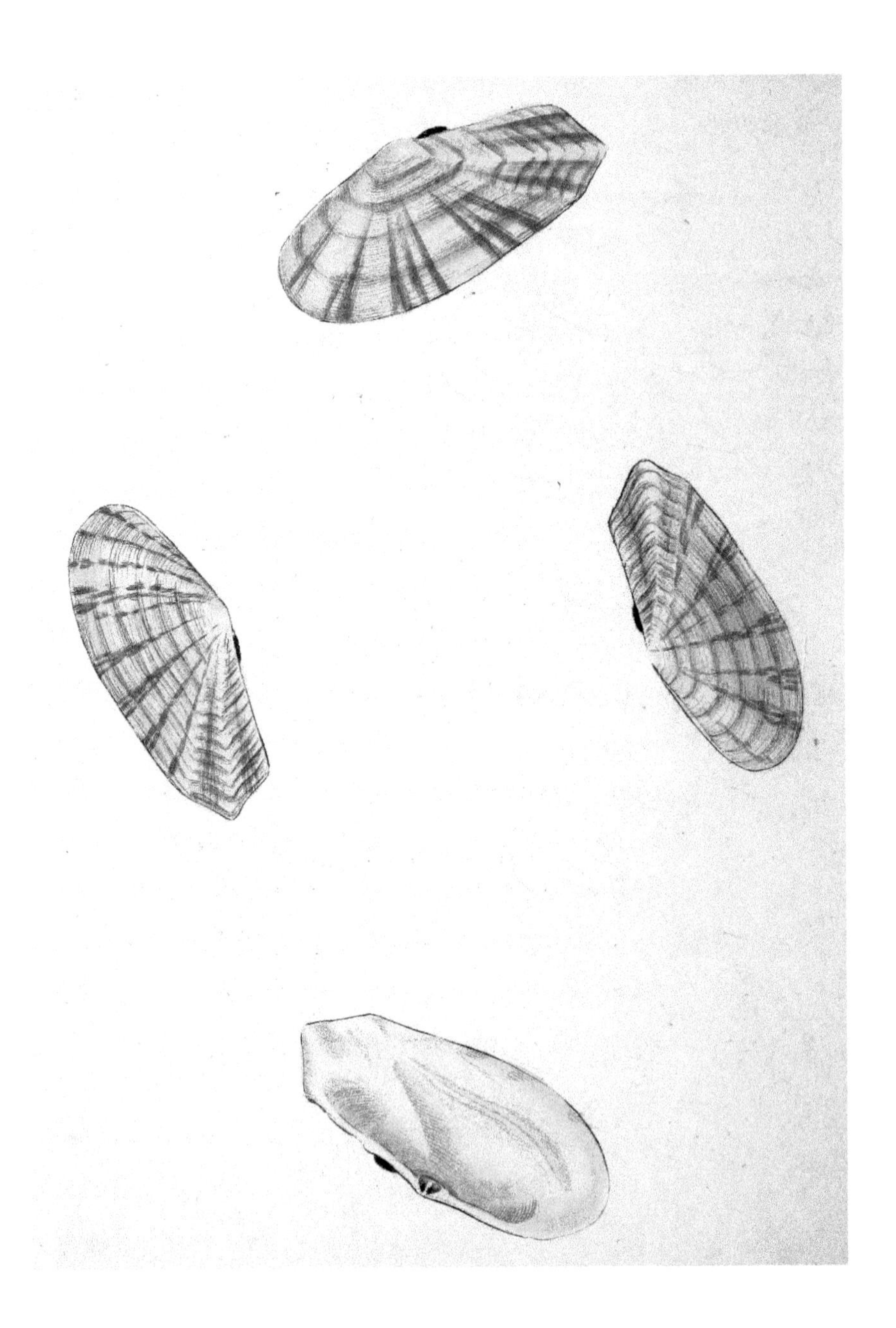

PLATE LX.

TELLINA TRIFASCIATA.

THREE STREAK TELLEN.

GENERIC CHARACTER.

The hinge usually furnished with three teeth; shell generally sloping on one side.

SPECIFIC CHARACTER.

Shell narrow oval, depressed, whitish, radiated with red; and striated transversely.

TELLINA TRIFASCIATA: testa ovata læviuscula sanguineo triradiata, pube rugosa. *Gmel. Linn. Syst. Nat. p.* 3233.

Tellina ex rufo maculata, fasciis exasperata. *List. H. Conch. tab.* 394. *fig.* 241.

Concha rugosa, tellinæ formis, lineola quadam paululum eminente ab ipso cardine ad imum ambitum donata. *List. App. Hist. An. Angl. p.* 19. *tit.* 36. *tab.* 1. *fig.* 8.—*App. Hist. An. Angl. in Gœdart. p.* 32. *tit.* 36. *tab.* 1. *fig.* 8.

Tellina cuneata compressa, e rubro radiata. Red Waved Bastard Tellen. *Petiv. Gazoph. tab.* 94. *fig.* 9.

Tellina depressa transversim striata albescens e rubro radiata, Radiata. *Da Costa. Br. Conch. p.* 209. *sp.* 42. *tab.* 14. *fig.* 1.

Tellina incarnata, carnation. *Penn. Br. Zool. No.* 31. *tab.* 47 *fig.* 31.

PLATE LX.

This elegant species is rather uncommon upon our coasts. Da Costa says he received it from Scarborough, in Yorkshire ; and adds, it is scarce on the coasts of Cornwall, but of a larger size ; the finest coloured specimens we have seen are from Dorsetshire and Wales.

The Tellina incarnata is smaller than trifaciata but very similar, and may be easily confounded with it.

61

PLATE LXI.

MACTRA SOLIDA.

GIRDLED.

GENERIC CHARACTER.

Animal a Tethys, Bivalve, sides unequal. Middle tooth complicated, with a little groove on each side ; lateral teeth remote.

SPECIFIC CHARACTER

AND

SYNONYMS.

Thick, transversely striated and girdled.

MACTRA SOLIDA : testa opaca læviuscula subantiqua. *Gmel. Linn. Syst. Nat. p.* 3259. *sp.* 13.

Concha crassa, ex altera parte compressa, ex altera subrotunda. *List. H. An. Angl. p.* 174. *tit.* 24. *tab.* 4. *fig.* 24.—Pectunculus crassiusculus albidus. *List. H. Conch. tab.* 253. *fig.* 87.

Chama media fasciata crassa *Petiv. Gaz. tab.* 94. *fig.* 7.

Chama minor plurimis fasciis. *Id. ib. fig.* 6.

A Pectunculus with azurine circular lines interpolated. *Leigh. Lancashire. tab.* 3. *fig.* 6.

Thick white striated Chama. *Wallis Northum. p.* 395.

PLATE LXI.

Mactra solida; strong. *Penn. Br. Zool. No.* 43. *tab.* 51. *fig.* 43. *A. et tab.* 52. *fig.* 43.—*Chemnitz. Conch.* 6. *t.* 23. *f.* 229. 230.

Trigonella crassa transversim fasciata. ZONARIA. *Da Costa. Br. Conch. tab.* 15. *fig.* 1. 1.

This species is found on many of our shores, as Kent, Dorsetshire, Lancashire, Yorkshire, Northumberland, the coast of Wales, &c.

The girdles are most prominent in the dead shells; the surface between them appearing much worn.

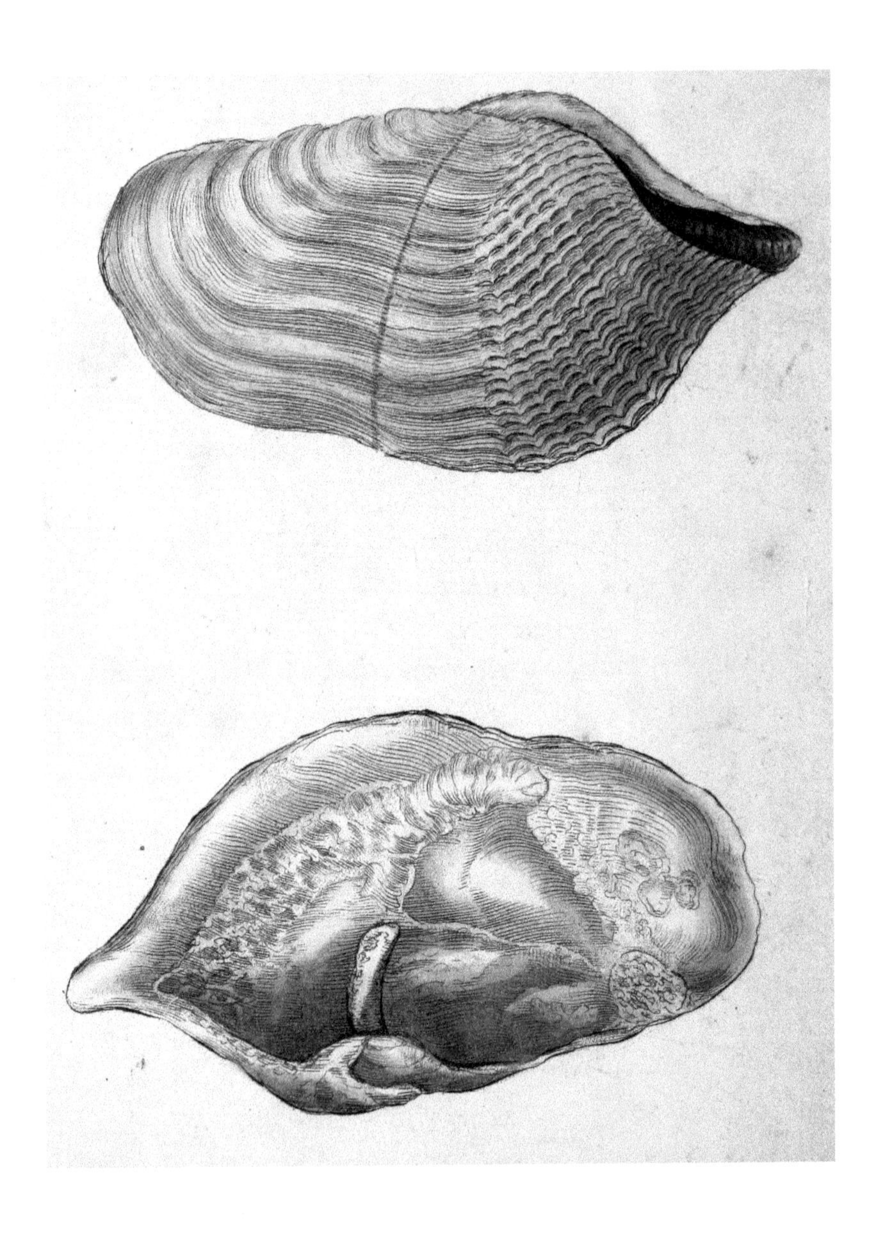

PLATE LXII.

PHOLAS CRISPATA.

CURLED, OR DOUBLE-FRONTED PIDDOCK.

GENERIC CHARACTER.

Animal Ascidia. Shell bivalve, opening wide at each end, with several lesser valves at the hinge. The hinges folded back and connected by a cartilage.

SPECIFIC CHARACTER
AND
SYNONYMS.

Shell oval, thick, wrought with transverse wrinkles, and divided down the middle by a furrow. The half next the hinge undulated or indented. A large flat curved tooth in the cavity under the beak.

PHOLAS CRISPATA: testa ovali hinc obtusiore crispato striata cardinis dente curvo. *Linn. Faun. Suec.* 2125. *Gmel. Linn. Syst. p.* 3216. *sp.* 6.

Concha altera parte dimidia striis undatim crispatis donata, altera lævis, apophysi longâ, angusta, recurva, dentiformi. An é peloridibus antiquorum? *List. H. An. Angl. p.* 192. *tit.* 38. *tab.* 5. 38.—Pholas angulosus, nobis olim, concha altera, &c. *Tit.* 38.—*App. H. An. Angl. in Goedart. p.* 36. *tab.* 2. *fig.* 7.—Pholas latus rugosus ex dimidio dorso et asper. *Hist. Conch. tab.* 279. *fig.* 436.

PLATE LXII

Concha ex dimidia pene margine profunde striata. *Merret Pin. p.* 194.

Chamæ pholas bifrons. Furrow-ribbed Pholas Muscle. *Petiver Gaz. tab.* 79. *fig.* 13.

Pholas Crispatus. Curled. *Penn. Br. Zool. No.* 12. *tab.* 40. *fig.* 12.

Pitaut, ou Dail Pholade. *Argenv. Conch.* 1. *p.* 365. *pl.* 30. *p.* 322. 26. *H.*

Pholas ovalis, parte dimidia striis undatim crispatis, altera lævis; dens longus angustus curvus. Bifrons. *Da Costa Br. Conch. p.* 242. *tab.* 16. *fig.* 4. 4.

Found in great abundance on many of our shores, nitched or burrowed in the rocks and stones in Cornwall, Lincolnshire, Yorkshire, Wales, &c.

PLATE LXIII.

ARCA NUCLEUS.

SILVERY ARK.

GENERIC CHARACTER.

Animal Tethys? Shell bivalve, equivalve. Teeth of the hinge numerous, inserted between each other.

SPECIFIC CHARACTER AND SYNONYMS.

Small, somewhat triangular, smooth, silvery within. Hinge semicircular, beset with numerous plate-like teeth. Margin finely crenated.

ARCA NUCLEUS: testa oblique ovata læviuscula, natibus incurvis, margine crenulato, cardine arcuato. *Linn. Syst. Nat. p.* 1143. *No.* 184.

ARCA NUCLEUS: testa oblique ovata læviuscula: cardine triangulari. *Gmel. Linn. Syst. Nat. p.* 3314. *sp.* 38.

Tellina inæquilatera, margine interno minutissime dentato, sed prope cardinem denticulis spissis, elatoribus, acutis, conspicua, oleagina, intus argentea. *Gualt* 1. *Conch. tab.* 81. *fig. R.*

G 2

PLATE LXIII.

Pectunculus minimus lævis, intus argenteus, cardine serrato. Silver Cockle. *Mus. Petiv. p.* 87. *No.* 841. *et Gazop. tab.* 17. *fig.*

Glycemeris Argentea parva subtriangularis, lævis, intus argentea. *Da Costa. Br. Conch. p.* 170. *sp.* 13. *tab.* 15. *fig.* 6. *right hand.*

This kind is found in great abundance on many of our shores, as Kent, Essex, Sussex, Devonshire, &c. and is also met with at Scarborough.

When these shells are fresh and perfect, says Da Costa, the outside is of an olive green, with some few transverse wrinkles; but when rubbed or worn are quite white, and almost smooth. The inside is of a fine silvery colour.

64

PLATE LXIV.

FIG. I.

TELLINA PLANA.

FLAT TELLEN.

GENERIC CHARACTER.

The hinge usually furnished with three teeth. Shell generally sloping on one side.

SPECIFIC CHARACTER

AND

SYNONYMS.

Somewhat triangular, thin, and flat.

TELLINA PLANA: tenuis subrotunda plana.

TRIGONELLA PLANA. *Da Costa. p.* 200. *sp.* 36.

Tellina crassa. Flat. *No.* 28.—Venus borealis. *Northern. Pen. Br. Zool. No.* 52.

Venus borealis. *Linn. Syst. Nat.?*

Concha tenuis, subrotunda, omnium minime cava, cardinis medio sinu et amplo et pyriformi. *List. H. An. Angl. p.* 174. *tit.* 23. *tab.* 4. *fig.* 23.

Pectunculus latus, admodum planus, tenuis, albidus. *List. H. Conch. tab.* 253. *fig.* 88.

Slender Smooth Chama. *Wallis Northumb. p.* 395.

G 3

PLATE LXIV.

In referring this ambiguous Shell to the Tellina genus, we may incur censure, as it does not certainly possess every characteristic of a tellen, yet we conceive less impropriety in altering the genera than in retaining it as a trigonella.

This Shell has been admitted as the Venus borealis of Linnæus and from the Synonyms of Lister's figure, not without probability. We do not, however, think the Linnæan descriptions agree sufficiently with our Shell; it may be a variety of it, though we hesitate to admit it as such.

Pennant has described this Shell twice, the old Shell is Tellina Crassa, No. 28, and the young one Venus borealis, No. 52 of that author; he adds indeed " the *Tellina crassa* has the habit of Venus borealis, but its sides are unequal, one being more extended than the other."

Da Costa has been under similar difficulties, he gives it as a species of his genus trigonella, though he says in the general description, that " the hinge of this kind is of a *different structure from the* TRIGONELLÆ, for it consists of two minute, thin, plate-like, parallel teeth, aside of which is a large triangular cavity, and has no lateral teeth."

Common on several of the English shores.

PLATE LXIV.

FIG. II.

TELLINA RIVALIS.

RIVER TELLEN.

SPECIFIC CHARACTER.

Shell oblique, somewhat ovated, furrowed transversely, and of an horn colour.

TELLINA RIVALIS : testa oblique subovata transversim sulcata cornei coloris. *Maton. Linn. Trans. vol.* 3.

The English naturalist is indebted to Dr. William Maton, author of the Tour of the Western Counties, for the discovery of this new and interesting British species. The first account of it appeared in a paper presented by him to the Linnæan Society, and afterwards inserted in their Transactions; and it is to this gentleman also our thanks are due for the specimens figured in the annexed plate. We have seen it since in the Collection of William Pilkington, Esq. Whitehall; who recently found it in the river near Hungerford in Berkshire.

Dr. Maton, in his remarks on this species, says, " It does not appear to have been described, and probably was never seen by Linnæus, nor has it been noticed by any English writer on Conchology ; a figure, however, of it occurs in *Gualteri's Index, Testacrum. Conchyliorum* (*Tab.* 7. *fig. C. C.*) but has been referred to by Professor

G 4

Gmelin, in his edition of the Systema Naturæ of Linnæus for *Tellina cornea*, though it evidently differs from the latter in shape, which Linnæus considers as one of the most certain *criteria*, whereby species are to be distinguished. *Gualtieri* mentions the Shells alluded to as "*Musculus fluviatilis, striatus, subflavus pellucidus*," which is a vague and imperfect description, and by no means sufficient to shew in what respect it differs from T. cornea." The difference consists chiefly in the T. rivalis being of a more oblique and subovated form, and in having the hinge near one end; T. cornea is somewhat globose, and in particular has the hinge and beaks placed in a more central manner.

Dr. Maton has generally found Tellina rivalis on chalky parts of the bed of the river Avon, and in rivulets communicating with it near Salisbury; but has never seen it in any considerable abundance. He conceives, that if diligently sought after, it may be discovered in most rivers and streams which are inhabited by *Tellina cornea*.

65

PLATE LXV.

HELIX ZONARIA.

STRIPED SNAIL.

GENERIC CHARACTER.

Aperture of the mouth contracted, and lunulated.

SPECIFIC CHARACTER

AND

SYNONYMS.

Shell whitish, striped, convex, rather depressed. A deep round central umbilicus. Outer lip of the mouth turned backward and spread.

HELIX ZONARIA: testa umbilicata convexa depressiuscula: apertura oblongiuscula marginata. *Linn. Syst. Nat. p.* 1245. *No.* 681.—*Gmel. Linn. Syst. Nat.* 3632. *sp.* 63. *Gualt.* 1. *Conch. tab.* 3. *fig. L.L.L.*

Cochleola alba fasciata cantabrigiensis, umbilico parvo. Newmarket Heath Shell. *Petiv. Gaz. tab.* 17. *fig.* 6.

Cochlea alba leviter umbilicata pluribus fasciis circumdata, clavicula productiore. *List. H. Conch. tab.* 59. *fig.* 56.

Cochlea umbilicata alba virgata. Virgata. *Da Costa. Br. Conch. p.* 79. *tab.* 4. *fig.* 7

PLATE LXV.

The Shells figured in the annexed plate are the true C. virgata of Da Costa, but not the Helix Zonaria of Pennant, as that author has erroneously considered them in his British Conchology. It appears that the latter species came into the possession of Da Costa after the work was published, for it stands corrected in some Mss. notes in his collection, though it is not noticed in his publication. Gmelin in his Systema Natura admits Da Costa's Shell as the Linnæan Zonaria; Pennant's Shell is not described by either author.

It inhabits dry sandy soils and banks, and, as Da Costa observes, is common only in some parts, as in the grass on Heddington-heath in Oxfordshire, and in Hampshire in plenty. It is also found in Cornwall, and was met with by Petiver on Newmarket-heath in Cambridgeshire.

66

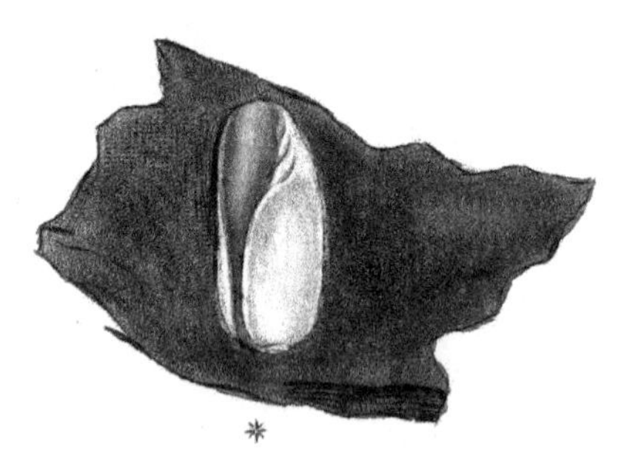

PLATE LXVI.

BULLA PALLIDA.

PALE, OR CYLINDRIC BULLA.

GENERIC CHARACTER.

Shell suboval. Aperture oblong, very patulous, and smooth or even. One end rather convoluted.

SPECIFIC CHARACTER.

Cylindric, white, glossy, four prominent wrinkles on the pillar lip.

BULLA PALLIDA: testa cylindrica, spira elevata acuta. *Linn. Mus. Reg. p.* 588. *No.* 223.

Voluta pallida testa integra oblongo ovata, spira elevata columella quadruplicata. *Linn. Syst. Nat. p.* 1189. *No.* 405.

Concha veneris, exigua, alba, vere cylindracea. *List. H. An. Angl. tab.* 714. *fig.* 70.

Porcellana integra admodum tenuis, fimbriata; dorso pulvinato, candidissima. *Gualt.* 1. *Conch. tab.* 15. *fig.* 4.

Bulla, cylindracea, cylindric. *Penn. Br. Zool. No.* 85. *tab.* 70. *fig.* 85.

Bulla exigua cylindracea, lævis et nivea. *Da Costa Br. Conch. p.* 30. *sp.* 16. *tab.* 2. *fig.* 7.

PLATE LXVI.

Bulla cylindricea is esteemed a very rare species by collectors of English Shells. It is found on the western coasts of England. Da Costa received them from Cornwall and Weymouth; and Lister notes them from Barnstaple in Devonshire.

The smallest figures denote the natural size.

67

PLATE LXVII.

VENUS DECUSSATA.

RETICULATED.

GENERIC CHARACTER.

Bivalve. Hinge furnished with three teeth ; two near each other, the third divergent from the beaks.

SPECIFIC CHARACTER

AND

SYNONYMS.

Somewhat oval, wrought with transverse and longitudinal striæ, or prominent ridges, which cross or decussate each other; outside brown, inside white, with violet spots near the hinge.

Cuneus reticulatus, longitudinaliter et transversim vel decussatim striatus, subrufus, intus ex parte violaceus. Reticulatus. Reticulated Purr. *Da Costa Br. Conch. p.* 202. *tab.* 14. *fig.* 4. 4.

VENUS DECUSSATA : testa ovata antice angulata decussatim striata. *Linn. Syst. Nat. p.* 1133. *No.* 149. *Mus. Reg. p.* 509. *No.* 77 ?

Concha quasi rhomboides, in medio cardine utrinque circiter tribus exiguis denticulatis donata. *List. H. An. Angl. p.* 171. *tit.* 20. *tab.* 4. *fig.* 20.

PLATE LXVII.

Chama fusca striis tenuissimis donata. *List. Hist. Conch. tab.* 423. *fig.* 271.*

Chama Purrs anglice dicta, et Tellina fasciata compactilis radiata intus ex parte subaurea, interdum subpurpurea. *List. Exercit. Anat.* 3. *p.* 25. 27. *tab.* 3.—*Wallace Orkn. p.* 42.—Chama nostras striis capillaceis. *Mus. Petiv. p.* 83. *No.* 811.

Purra fasciata et radiata. Cornwall Purr. *Petiv. Gaz. tab.* 95. *fig.* 8.—Chama, Purrs. *Dale Harw. p.* 387. *No.* 5.

Venus litterata, lettered. *Penn. Brit. Zool. p.* 96. 53.

The young Shells of this species vary considerably in their colours and markings, but are in general remarkable for their elegance; as they encrease in growth, those colours and markings gradually fade, and in old Shells become altogether obscure. It is found in plenty on most of the southern coasts of England and Wales.

---

* Gmelin makes a new species of Lister's shell in the Systema Naturæ, under the name *obscura*, without noticing any other author who describes the same kind "*Venus obscura* testa fusca: striis perpendicularibus tenuissimis, p. 3289. sp. 99."

68

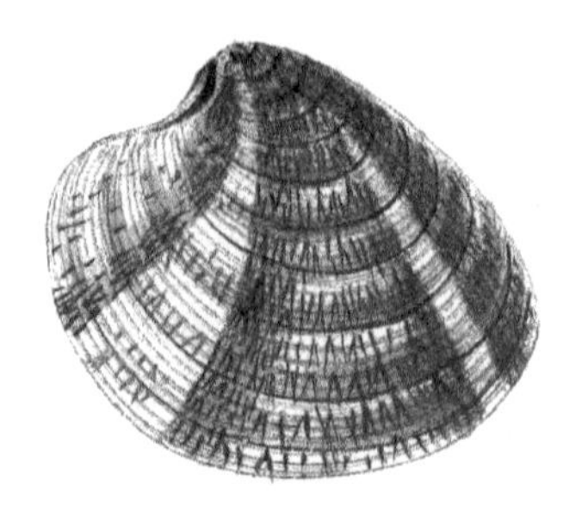

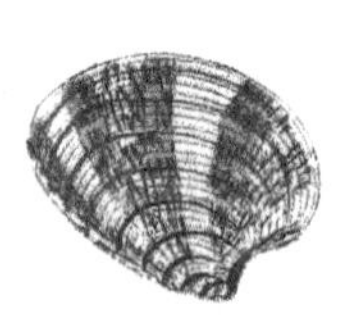

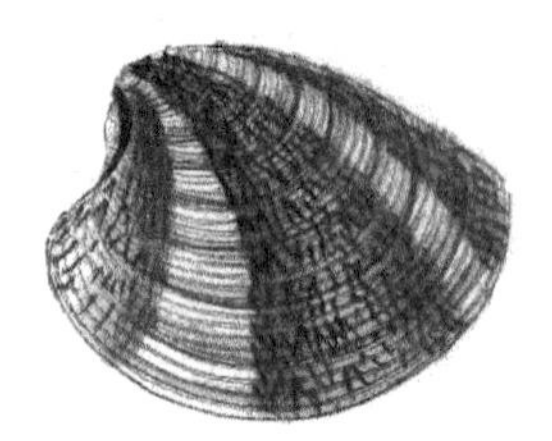

PLATE LXVIII.

VENUS STRIATULUS.

STRIATED.

GENERIC CHARACTER.

Bivalve. Hinge furnished with three teeth ; two near each other, the third divergent from the beaks.

SPECIFIC CHARACTER.

Shell somewhat heart-shaped, and marked with three or four longitudinal rays of brown.

PECTUNCULUS STRIATULUS parvus transversim striatus fusco radiatus. *Da Costa. Br. Conch. p.* 191. *sp.* 27. *tab.* 12. *fig.* 2. 2.

Venus Gallina testa subcordata radiata : striis transversis obtusis, cardinis dente postico minimo, margine crenulato. *Linn. Syst. Nat. p.* 1130. *No.* 119.—*Fn. Sv.* 2. *No.* 2143?

This is one of the most elegant of the British Shells. It is found on the coasts of Dorsetshire, Cornwall, and the isles of Scilly, and also on those of Wales. The general colour is pale flesh colour, radiated and figured with a chestnut brown, but in some instances they vary to an uniform brown or orange, obscurely spotted with black.

Da Costa is the only English author who notices this species.

PLATE LXIX.

PHOLAS PARVUS.

SMALL PIDDOCK.

GENERIC CHARACTER.

Animal ascidia. Shell bivalve, opening wide at each end, with several lesser valves at the hinge. The hinges folded back and connected by a cartilage.

SPECIFIC CHARACTER

AND

SYNONYMS.

Shell oval, thin, wrought with transverse wrinkles, and divided down the middle by a furrow. The half next the hinge undulated or indented. A slender and oblique curved tooth in the cavity under the beak. Size of a hazel nut.

Pholas Parvus. Little. *Pen. Br. Zool. sp.* 13.
PHOLAS PARVUS: simillima tota structura Pholade Bifronte. *Da Costa Br. Conch. p.* 247. *sp.* 67.

This shell was first described by Pennant in his Zoology; he says it very much resembles the Pholas crispatus but is never found larger than a hazel nut. Da Costa describes it also, but doubts whether it is a diſtinct species or only a young shell of that kind. As both authors have however figured and described it separately, we have given it a place as a distinct, or at leaſt doubtful species.

Pennant says he found these shells in masses of fossil wood in the shores of Abergelli in Denbighshire: the bottom of their cells were round and appeared as if nicely turned with some instrument. According to this author they will also perforate the hardest oak plank that is accidentally lodged in the water. Da Costa says they are found in great quantities on the same coasts as the other kind (Pholas crispatus) nitched in the rocks and stones, and adds that there is an amazing abundance at Scarborough and Whitby in Yorkshire, nitched in the Alum and other stones.

70

PLATE LXX.

MYTILUS BARBATUS.

BEARDED MUSCLE.

GENERIC CHARACTER.

The hinge toothless, and consists of a longitudinal furrow.

SPECIFIC CHARACTER

AND

SYNONYMS.

Short, ventricose, obtuse, ferruginous yellow. An oblique space extending from the hinge to the apex, covered with a rude epidermis and irregular filaments.

MYTILUS BARBATUS: testa læviuscula ferruginea exterius apice barbata. *Fn. Suec.* 2157. *Gmel. Linn. Syst. Nat. p.* 3353. *sp.* 10. *Chemn. Conch.* 8. *t.* 84. *f.* 749.

In trawling for marine productions in the Straits that divides Caernarvon from Anglesea (Menai) last summer, we found several specimens of the Mytilus barbatus of a much larger size than any hitherto described by authors, as the figures in the annexed plate will fully express.

Da Costa has not noticed this species, though it must have been known to him from the figures and descriptions in Pennant's Zoology, where it stands under the name of M. Curtus. *sp.* 76. *A.* Short. Pennant's specimen scarcely exceeds the size of the second specimen figured in our plate;—it was described from a Shell in the Portland Cabinet, that had been taken at Weymouth.

Linnæus mentions this species in the Fauna Suecica. Gmelin quotes the Works of Chemnitz for its figure, where it appears somewhat smaller than in those of Pennant. It is certainly very scarce.

7'

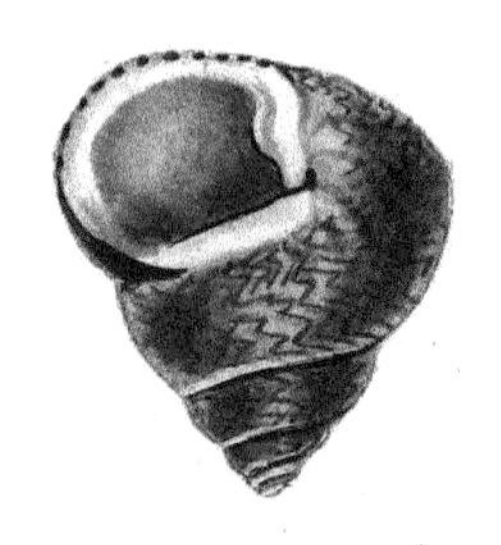

PLATE LXXI.

TURBO LINEATUS.

STREAKED.

GENERIC CHARACTER.

Animal Limax. Univalve, spiral, or of a taper form. Aperture somewhat compressed, orbicular, entire.

SPECIFIC CHARACTER.

Somewhat conic. Ash colour variegated with fine streaks and irregular marks of black; a rude tooth at the top of the pillar.

TURBO LINEATUS: trochiformis cinereus lineis aut lituris nigris insignitus, columella subdentata. *Da Costa Br. Conch. p.* 100. *sp.* 56. *tab.* 6. *fig.* 7.

We believe this species is rather an uncommon, or at least local kind on the British shores, though Da Costa says it is found on the coasts of Devonshire, Cornwall, Dorsetshire, Pwllhely in Caernarvonshire, and in plenty on the coasts of Norfolk. The collection of that author contains but a single specimen, it is a worn Shell and indifferently expressed by the figure above quoted. The most characteristic Shells of this species we have seen, we found on the rocky

H 3

shores of Aberfraw, on the western side of Anglesea, and at Manachty the remotest part of the same island.

This Shell is large, thick, and conic or shaped like a trochus. The general colour is ashen with little variation, the lines in some are dark or almost black, in others of a pale brown, or brown tinged with red; when the external covering is worn off the Shell appears of a fine mother of pearl.

Turbo lineatus is not described by any English Author except Da Costa.

72

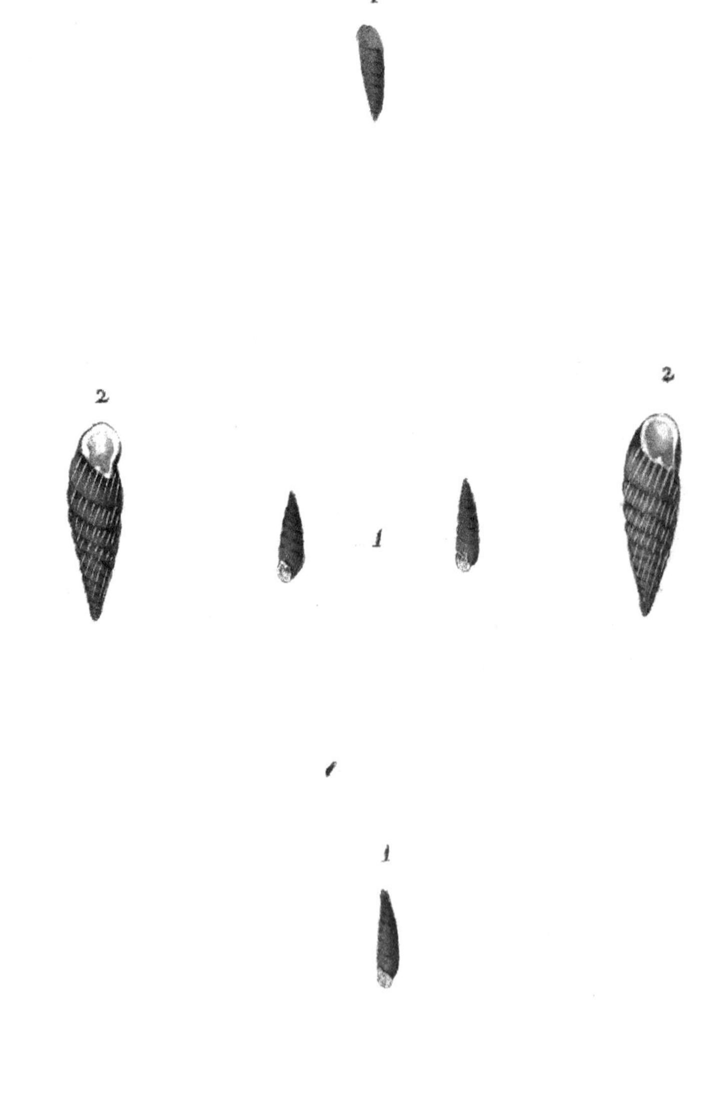

PLATE LXXII.

TURBO PERVERSUS.

REVERSED, OR OAT.

GENERIC CHARACTER.

Animal Limax. Univalve, spiral, or of a taper form. Aperture somewhat compressed, orbicular, entire.

SPECIFIC CHARACTER.

Taper, somewhat transparent. Spires turn from left to right. Mouth jagged or beset with teeth.

TURBO PERVERSUS: testa turrita pellucida, anfractibus contrariis apertura edentula. *Linn. Syst. Nat. p.* 1240. *No.* 650.

Cochlea testa pellucida oblonga, spiris decem sinistrorsis, apertura subrotunda. *Linn. Faun. Suec.* 1. *p.* 372. *No.* 1300. 2. *No.* 2172.

Buccinum pullum, opacum, ore compresso, circiter denis spiris fastigiatum. *List. H. An. Angl. p.* 123. *tit.* 10. *tab.* 2. *fig.* 10.

Buccinum exiguum pullum duodecim orbium. *List. H. Conch. tab.* 41. *fig.* 39. *Maj. et. min.*

H 4

PLATE LXXII.

Buccinum alterum pellucidum subflavum, intra senas circiter spiras mucronatum. *List. H. An. Angl. p.* 124. *tit.* 11. *tab.* 2. *fig.* 11.—*Phil. trans. No.* 105. *fig.* 11.

The small Whirl Snail, with numerous rounds, and winding from the mouth towards the right hand. *Grew. Mus. p.* 132.—*Morton Northampt. p.* 415.—Et Buccinum heterostrophum minutum fuscum sex spirarum ore subrotundo. *Id. p.* 416. *tab.* 13. *fig.* 1.

Buccinulum Anglicum heterostrophon oblongum striis capillaceis. *Petiv. Mus. p.* 65. *No.* 703.

Turbo perversus. Reversed. *Penn. Br. Zool. No.* 116. *tab.* 82. *fig.* 116.

Strombiformis parvus pullus, ore compresso, anfractibus contrariis striatis. PERVERSUS, Reversed Oat. *Da Costa Br. Conch. p.* 107. *tab.* 5. *fig.* 15. 15.

This is one of the heterostrope Shells, or such as have the mouth placed on the right side instead of the left, as is usual in most species. In general, heterosphe Shells are mere accidental varieties only of such as turn in the usual manner* ; but in the present instance, it constitutes a striking character of the species itself.

It is a matter of some difficulty to reconcile the various opinions of authors respecting the several varieties and growths of this species; Da Costa has entered into the enquiry; and the result of his remarks appear at least satisfactory to us.

---

* Reversed shells of the common garden snail have been found, though very rarely. One is mentioned by Dr. Latham in his Synopsis of Birds.

PLATE LXXII.

" These smaller ones," says Da Costa, " are the young Shells, but always with them are found old ones of double or treble the size; in every other respect like these, but proportionally larger and stronger in their several parts and work. The plaits or foldings near the mouth are deep and very strong; the striæ stronger and more distinct; the border round the mouth greatly turned outwards, very broad, flat, thick, milk white, and the sinuosities, jags or teeth, within, are large, white, and very conspicuous; some are bidentated, and most of these old ones have eleven, and some even twelve spires.

" From these circumstances, authors run into confusion, by making the different growths different species. The accurate and judicious Lister himself has formed two species, in his tit. 10. and 11. on the difference of the number of the spires and other slight particulars. The several figures in Gualtieri are only varieties; and the bidens of Linné, Syst. Nat. p. 1240. No. 649. and of Mr. Pennant, Brit. Zool. No. 117. tab. 81. fig. 117. is apparently no other than an old Shell, for such large and bidentated ones I have not unfrequently found nestled with these common smaller Shells.

" Though the number of spires in a Shell is a criterion, yet it is not an infallible one, for the number of spires vary in some species, either from the growths or sexes: in such cases the young Shells have always a less number, and the males have their spires less numerous than the females. This very species is, perhaps, as strong an instance of the difference in the number of the spires as can be, for it is found from six to twelve spires, as Linné has also noted in his Fauna Suecica."

Linnæus, and Gmelin in his last Systema Naturæ, distinguish the two species Bidens and Perversus chiefly by the number of teeth. The latter is described with three teeth, the former of course with only two. Fig. 1. 1. denotes the natural size. Fig 2. magnified.

INDEX.

VOL. II.

LINNÆAN ARRANGEMENT.

MULTIVALVIA.

BIVALVIÆ. CONCHA.

INDEX.

UNIVALVIA.

INDEX TO VOL. II.

ACCORDING TO

HISTORIA NATURALIS TESTACEORUM BRITANNIÆ OF DA COSTA.

PART I.

GENUS 4.

* MARINÆ. SEA.

PART II.

UNIVALVIA INVOLUTA.

GENUS 5. BULLA. DIPPER.

GENUS 6. CYPRÆA, COWRY.

INDEX.

PART III.

UNIVALVIA TURBINATA.

TROCHUS TOP SHELL.

* MARINÆ. SEA.

GENUS 9. HELIX

FLUVIATILES. RIVER.

GENUS 34.

COCHLEA SNAILS.

TERRESTRES. LAND.

GENUS 41.

TERRESTRES. LAND.

TURBO.

I N D E X.

FLUVIATILES. RIVER.

MARINÆ. SEA.

GENUS 12. STROMBIFORMIS. NEEDLE SNAIL.

TERRESTRES. LAND.

MARINÆ. SEA.

ORDER 2.

BIVALVES.

GENUS 1. PECTEN. ESCALLOP.

GENUS 2. OSTREUM. OYSTER.

INDEX.

PART. II.

MARINÆ. SEA.

GENUS 4. GLYCYMERIS.

GENUS 6. CARDIUM. HEART COCKLE.

MARINÆ. SEA.

PECTUNCULUS. COCKLE.

GENUS 3. TRIGONELLA.

MARINÆ SEA.

GENUS 9. CUNEUS. PURR.

GENUS 10. TELLINA. TELLEN.

INDEX.

GENUS 11. MYTILUS MUSCLE.

FLUVIATILES. RIVER.

MARINÆ. SEA.

PART III.

GENUS 13. CHAMA. GAPER.

MARINÆ. SEA.

GENUS 14. SOLEN. SHEATH OR RAZOR SHELL.

PART IV.

MULTIVALVES.

GENUS 16. PHOLAS. PIDDOCKS.

I N D E X.

GENUS 17. BALANUS. ACORN.

MARINÆ. SEA.

ALPHABETICAL INDEX TO VOL. II.

	Plate.	Fig.
ACUTA Helix, Sharp - - - -	39	2
Auricularia Helix, Ear, or Wide Mouth River Snail -	51	1
Barbatus Mytilus, Bearded - - - -	70	
Borealis, Tellina - - - - -	62	
Carnaria, Tellina, Flesh coloured Tellen - -	47	
Cornea, Helix, Ram's Horn - - -	39	1
Corneus, Murex, Horny or slender Whelk - -	38	
Crispata, Pholas, Curled or Double fronted Piddock -	62	
Cygnæus, Mytilus, Great Horse or Swan Muscle -	55	
Decussata, Venus, reticulated - - -	67	
Diadema, Lepas, Whale Acorn Shell - - -	56	
Ensis, Solen, Scymetar - - - -	50	
Entalis, Dentalium, Tooth Shell - - -	48	
Exoleta, Venus, antiquated - - - -	42	1
Glycymeris, Arca, Orbicular Ark - - -	37	
Inæquivalvis, Tellina, Unequal-valved Tellen - -	41	
Lapicida, Helix, Acute-edged - - -	39	2
Lævigatum, Cardium, Large High-beaked Cockle -	54	
Lægumen, Solen, Peasecod - - - -	53	
Lineatus, Turbo, streaked - - - -	71	
Lutraria, Mactra, Large Gaper - - -	58	
Maxima, Ostrea, Great Scallop - - -	49	
Nucleus, Arca, Silvery Ark - - - -	63	
Pallida, Bulla, Pale or Cylindric Bulla - - -	66	
Parvus, Pholas, Small Piddock - - -	69	
Pediculus, Cypræa, Sea Louse, Cowry, or Nun - -	43	
Perversus, Turbo, Reversed or Oat - - -	72	
Rivalis, Tellina - - - - -	62	2
Siliqua, Solen, Large or Pod Solen - - -	46	
Sinuosa, Venus, Indented Venus Shell - - -	42	2
Solida, Mactra, Girdled - - - -	61	
Stagnalis, Helix, Lake Snail - - - -	51	2

INDEX.

	Plate.	Fig.
Striata, Ostrea, Striated Oyster - - -	45	
Striatulus, Venus, striated - - - -	68	
Striatus, Turbo, striated Wreath Shell - - -	59	
Tornatilis, Voluta, Oval volute - - -	57	
Trifasciata, Tellina, Three-streaked Tellen - -	60	
Variabilis Tellina, variable - - - -	41	2
Verrucosa, Venus, Warted Venus Shell - -	44	
Umbilicatus, Mytilus, Umbilicated or Wry Beak Muscle -	40	
Zizyphinus, Trochus, Livid Top Shell - - -	52	
Zonaria, Helix, Striped Snail - - - -	65	

END OF VOL. II.

Printed by Bye and Law, St. John's-Square, Clerkenwell.

THE

NATURAL HISTORY

OF

BRITISH SHELLS,

INCLUDING

FIGURES AND DESCRIPTIONS

OF ALL THE

SPECIES HITHERTO DISCOVERED IN GREAT BRITAIN,

SYSTEMATICALLY ARRANGED

IN THE LINNEAN MANNER,

WITH

SCIENTIFIC AND GENERAL OBSERVATIONS ON EACH.

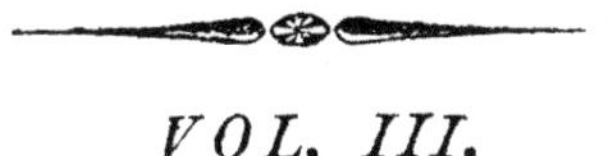

VOL. III.

BY E. DONOVAN, F.L.S.

AUTHOR OF THE NATURAL HISTORIES OF

BRITISH BIRDS, INSECTS, &c. &c.

LONDON:

PRINTED FOR THE AUTHOR,

AND FOR

F. AND C. RIVINGTON, No 62, ST. PAUL'S CHURCH-YARD.

BY BYE AND LAW, ST. JOHN'S SQUARE, CLERKENWELL.

1801.

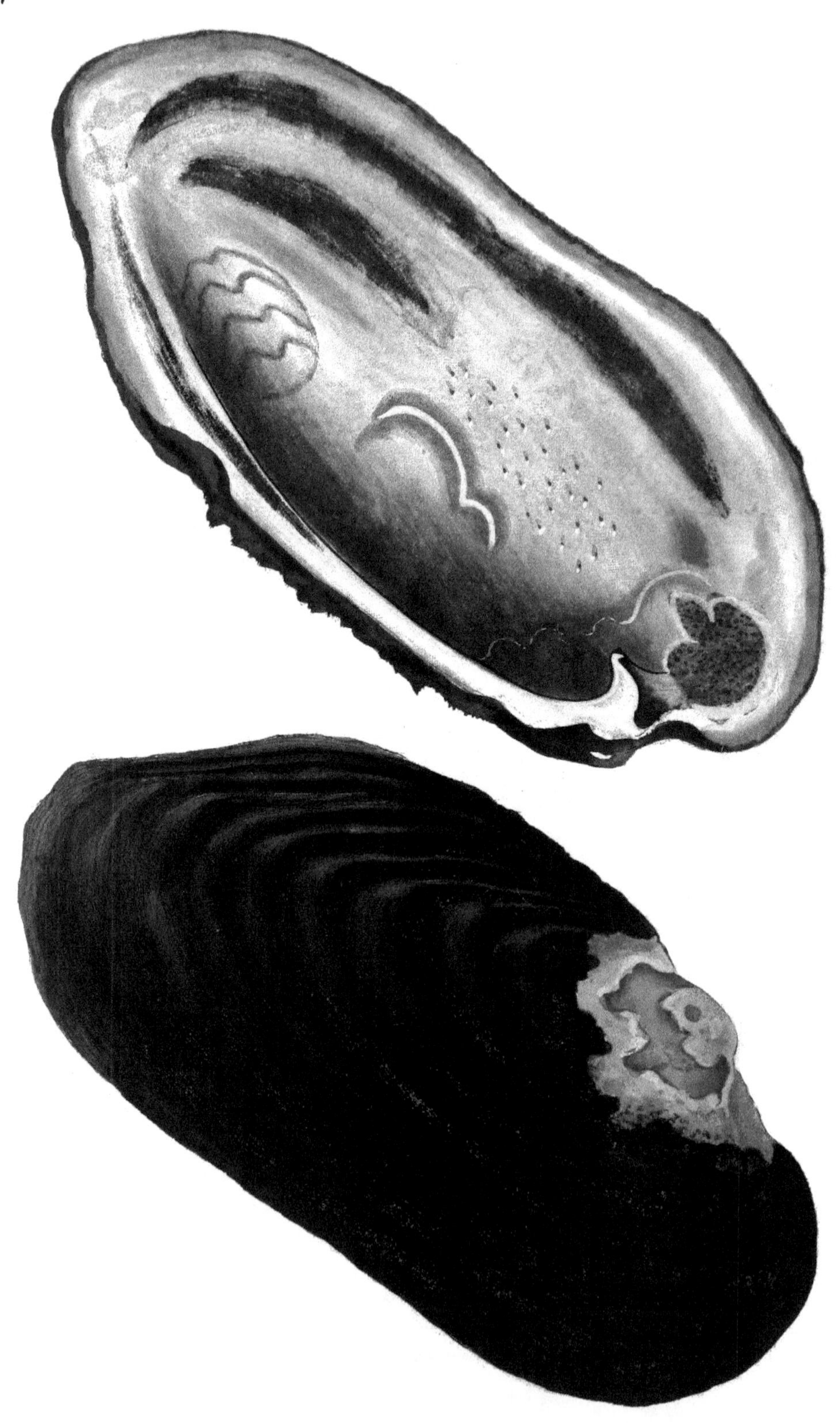
73

THE

NATURAL HISTORY

OF

BRITISH SHELLS.

PLATE LXXIII.

MYA MARGARITIFERA.

RIVER PEARL MUSCLE.

GENERIC CHARACTER.

Animal an Ascidia. Shell bivalve, gaping at one end. The hinge for the most part furnished with a thick strong broad tooth, not inserted into the oppofite valve.

SPECIFIC CHARACTER

AND

SYNONYMS.

Shell oblong, thick, and covered with a coarse black epidermis, much decorticated or worn down about the beaks. A large notched conic tooth in one valve, and two small ones in the other.

A 2

PLATE LXXIII.

MYA MARGARITIFERA: testa ovata anterius coarctata, cardinis dente primario conico, natibus decorticatis. *Linn. Fn. Suec.* 2130.—*Gmel. Linn. Syst. Nat.* 3219. *sp.* 4.

Mya nigrescens crassa & ponderosa margaritifera. Margaritifera. *Da Costa Br. Conch. p.* 225. *sp.* 53. *tab.* 15. *fig.* 3. 3.

Musculus niger omnium crassissima et ponderosissima testa. Conchæ longæ species. *Gesn. & Aldrov. List. App. H. An. Angl. p.* 11. *tit.* 31. *tab.* 1. *fig.* 1. & *App. H. An. Angl. in Goed. p.* 15. *tit.* 31. *tab.* 1. *fig.* 1.

Musculus niger omnium longe crassissimus. Conchæ longæ species. *Gesn. & Aldr. Hist. Conch. tab.* 149. *fig.* 4.

Musculi margaritiferi. *Bede Hist. Ecclesiast. I.* 1. *c.* 1.

Martin's West. Isles. p. 7. &c.

Pearl Muscles. Leigh Lancashire, p. 134.

Mytulus major margaritiferus. *Wallis Northumb. p.* 403. *No.* 42.

Mya margaritifera, Pearl. *Penn. Br. Zool. No.* 18. *tab.* 43. *fig.* 18.

" This shell," says Pennant, " is noted for producing quantities of pearl. There have been regular fisheries for the sake of this precious article in several of our rivers. Sixteen have been found in one shell. They are the disease of the fish analogous to the stone in the human body. On being squeezed, they will eject the pearl, and often cast it spontaneously in the sand of the stream.

" The Conway was noted for them in the days of Camden. A notion also prevails that Sir Richard Wynne, of Gwydir, chamberlain to Catherine queen to Charles II. presented her majesty with a pearl (taken in this river) which is to this day honoured with a place in the regal crown. They are called by the Welsh Cregin Diluw, or Deluge Shells, as if left there by the flood.

" The Irt, in Cumberland, was also productive of them. The famous circumnavigator, Sir John Hawkins, had a patent for fishing that river. He had observed pearls plentiful in the straits of Magellan, and flattered himself with being enriched by procuring them within his own island.

" In the last century, several of great size were gotten in the rivers in the county of Tyrone and Donegal, in Ireland. One weighed thirty-six *carats*, was valued at 40l. but being foul lost much of its worth. Other single pearls were sold for 4l. 10s. and even for 10l. The last was sold a second time to lady Glenlealy, who put it into a necklace, and refused 80l. for it from the duchess of Ormond."

" Suetonius reports, that Cæsar was induced to undertake his British expedition for the sake of our pearls; and that they were so large that it was necessary to use the hand to try the weight of a single one*. I imagine Cæsar only heard this by report; and that the crystaline balls in old leases, called mineral pearl, were mistaken for them †."

" We believe that Cæsar was disappointed of his hope: yet we are told that he brought home a buckler made with British pearl ‡, which

---

* *Sueton. Vit. Jul. Cæs.* c. lxiv. † Woodward's Method of Fossils, 29. part 2.

‡ *Plinii, lib.* 9. *c.* 35. *Tacit. Vit. Agricolæ.*

he dedicated to, and hung up in the temple of Venus Genetrix. A proper offering to the goddess of beauty, who sprung from the sea. I cannot omit mentioning, that notwithstanding the classics honour our pearl with their notice, yet they report them to be small and ill coloured; an imputation that in general they are still liable to. Pliny says, "that a red small kind was found about the Thracian Bosphorus, in a shell called Mya, but does not give it any mark to ascertain the species."

The Mya Margaritifera is found only in great rivers, and chiefly in those of the northern parts of Great Britain. Da Costa mentions the Tees, Alne, North and South Tyne, Tweed, Dee, Don, &c. and adds, generally inhabits the deeper parts, as gulphs, whirlpools, &c.

The fishermen in the neighbourhood of the river Conway sometimes collect those shells, and extract the pearl, but as they are now become scarce, and the price inconsiderable, the fishery affords them little encouragement.

74

PLATE LXXIV.

TROCHUS CINERARIUS.

UMBILICAL TOP SHELL.

GENERIC CHARACTER.

Animal a slug. Shell spiral sub-conic.

SPECIFIC CHARACTER

AND

SYNONYMS.

Umbilicated or perforated at *the base.* Not very conic. Five whirls.—Colours various, generally greenish, radiated obliquely with red or brown.

TROCHUS CINERARIUS: testa oblique umbilicata, ovata, anfractibus rotundatis. *Linn. Syst. Nat. p.* 1229. *No.* 590.

Trochus planior umbilicatus, undatim ex fusco perbelle radiatus, UMBILICALIS *Da Costa. Br. Conch. p.* 46. *tab.* 3. *fig.* 4. 4.

Trochus planior undatim ex rubro late radiatus. *List. H. Conch. tab.* 641. *fig.* 32.

Umbilicated Top shell. *Dale Harwich. p.* 381. *No.* 4.

Trochus Umbilicaris. Umbilical. *Penn. Br. Zool. No.* 106. *tab.* 80. *fig.* 106.

A very common species on most of the British shores.

A 4

75

PLATE LXXV.

HELIX VORTEX.

COMMON WHIRL SHELL.

GENERIC CHARACTER.

Aperture of the mouth contracted and lunulated.

SPECIFIC CHARACTER

AND

SYNONYMS.

Shell of five wreaths, horizontal. Somewhat convex on the upper side, under side flat, and carinated, or surrounded with a sharp edge.

HELIX VORTEX: testa carinata; supra concava, aperture ovali plana. *Linn. Syst. Nat. p.* 1243. *No.* 667.

Cochlea testa plana fusca: supra concava, anfractibus quinque, margine acuto. *Linn. Fn. Suec. I. p.* 374. *No.* 130. 7. 11. *No.* 2172.

Cochlea exigua, subfusca, altera parte planior, sine limbo, quinque spirarum. *List. H. An. Angl. p.* 145. *tit.* 28. *tab.* 2. *fig.* 28.

Cochlea exigua quinque orbium. *List. Conch. tab.* 138. *fig.* 43.

Planorbis polygirata minor. *Petiv. Gaz. tab.* 92. *fig.* 6. 7. *Morton Northampt. p.* 417.

Helix vortex. Whirl. *Penn. Br. Zool. No.* 124. *tab.* 83. *fig.* 124.

PLATE LXXV.

Cochlea exigua plana sine limbo. Planorbis. *Da Costa. Br. Conch. p.* 65. *sp.* 36. *tab.* 4. *fig.* 12.

A very common species of aquatic snail in stagnant waters and rivers. It is flat and thin, and has not a prominent border as in Helix Limbata of Da Costa, or Helix Planorbis of Pennant.

76

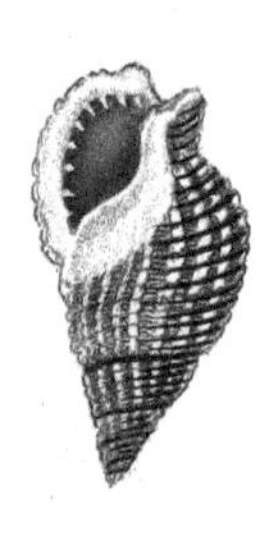

PLATE LXXVI.

BUCCINUM RETICULATUM.

RETICULATED WRY MOUTHED WHELK.

GENERIC CHARACTER.

Animal slug. Shell spiral, gibbous, aperture oval, ending in a short canal.

SPECIFIC CHARACTER

AND

SYNONYMS.

Oblong, reticulated, or furrowed transversely and longitudinally. Mouth beset with prominent teeth.

BUCCINUM RETICULATUM: testa ovato-oblonga transversim ftriata, longitudinaliter rugosa, apertura dentata. *Gmel. Linn. Syst. Nat. p.* 3495. *sp.* 111.

Buccinum recurvirostrum cancellatum, columella sinuosa, labro dentato. Reticulatum. *Da Cofta. Br. Conch. p.* 131. *tab.* 7. *fig.* 10.

Buccinum brevi rostrum cancellatum, dense sinuosum, labro dentato. *List. H. Conch. tab.* 966. *fig.* 21.

Buccinum marinum cancellatum. Small latticed Whelke. *Petiv. Gaz. tab.* 75. *fig.* 4.

PLATE LXXVI.

Dale Harw. p. 283. *No.* 7. & *p.* 285. *No.* 3.

Smooth chequered Whelk. *Smith. Cork. p.* 318.

Very common on several of our sea coasts, especially on those of Essex, Kent, Sussex, &c. Also found in Wales and Ireland.

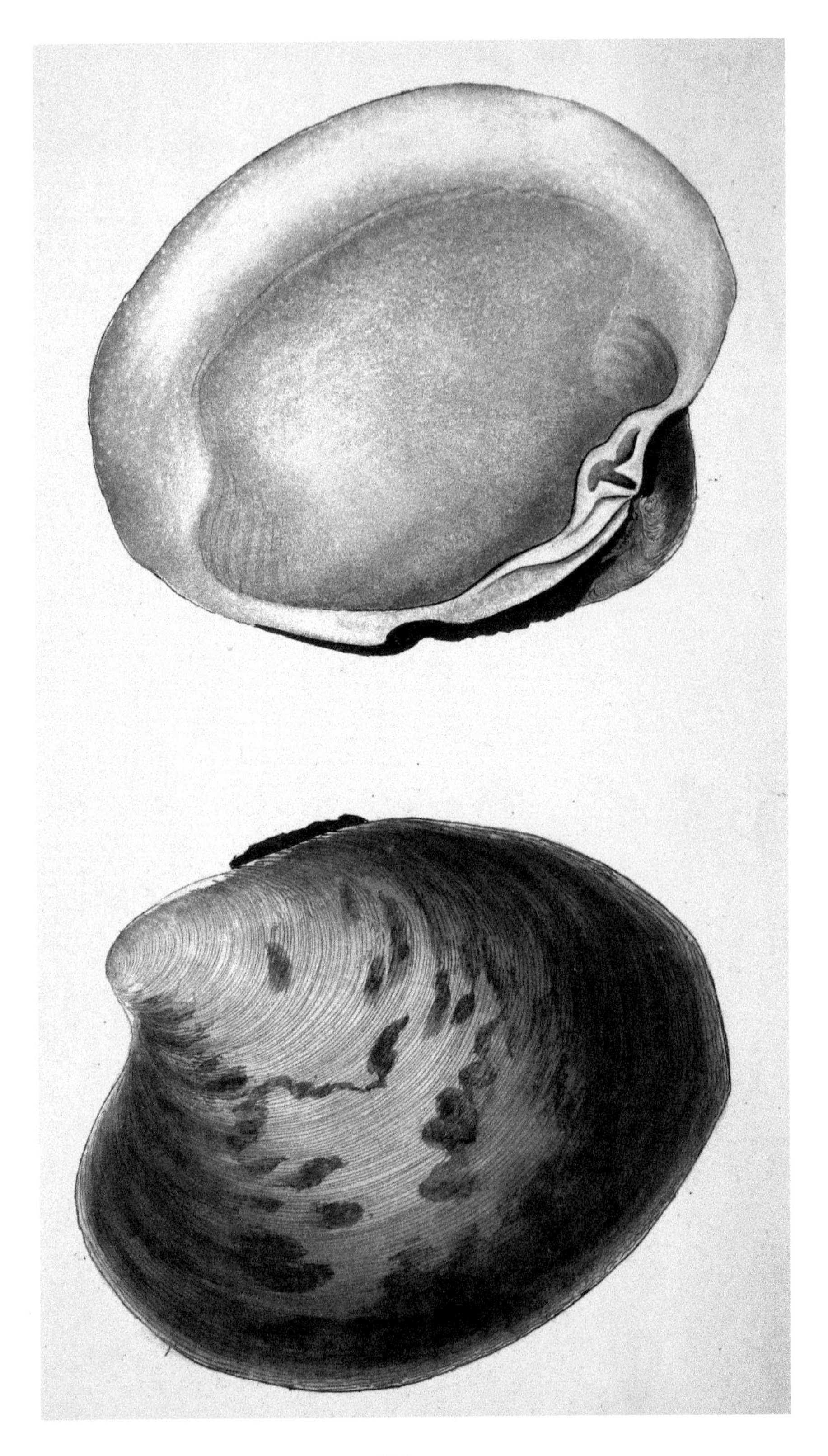

PLATE LXXVII.

VENUS ISLANDICA.

THICK VENUS.

GENERIC CHARACTER.

Bivalve. Hinge furnished with three teeth; two near each other, the third divergent from the beaks.

SPECIFIC CHARACTER

AND

SYNONYMS.

Shell strong, thick, heavy, covered with epidermis; space in which the hinge is inserted gaping. Margin acute and entire. White within.

VENUS ISLANDICA: testa transversim striata rudi, nymphis hiantibus, ano nullo. *Gmel. Linn. Syst. Nat.* 3271. *sp.* 15.

Pectunculus major crassus, albo castaneus. Crassus, *Da Costa Br. Conch. sp.* 183. *tab.* 14. *fig.* 5.

Concha é maximis, admodum Crassa, rotunda, ex nigro rufescens. *List, H. An. Angl. p.* 170. *tit.* 22. *tab.* 4. *fig.* 22.

Pectunculus maximus, subfuscus, valde gravis. *List. H. Conch. tab.* 272. *fig.* 108.

Venus mercenaria. Commercial. *Penn. Br. Zool. No.* 47. *tab.* 53. *fig.* 47.

PLATE LXXVII.

Chama inæquilatera, lævis, crassa, subalbida. *Gualt.* 1. *Conch. tab.* 85. *fig. B.*

Da Costa notices a material error amongst the synonyms Linnæus has given with his description of Venus Mercenaria. The Venus Mercenaria of Linnæus is the shell called North American Clam, and of which the Wampum, or indian money, is made; this is not the species found on our coast and figured by Lister, p. 173. as Linnæus imagined, but a shell altogether distinct; the English species Lister notices, is the true Venus Islandica of the Linnæan *Systema Naturæ.*

This error has misled Pennant, who confounds the North American kind with our species, at least as a variety having a purple tinge within it. Gmelin was aware of this mistake, for in his edition of the Systema Naturæ, both the plates and descriptions of Pennant and Da Costa are referred to in the synonyms of Venus Islandica.

This shell is perfectly white when fine, and is thickly covered with a fibrous epidermis of a black, or brownish colour. Found on several of our coasts. Da Costa mentions Northumberland, Yorkshire, Lancashire, Dorsetshire, Caernarvonshire, and other shores of Wales. Aberdeenshire, and the islands of Orkney, &c, in Scotland.

78

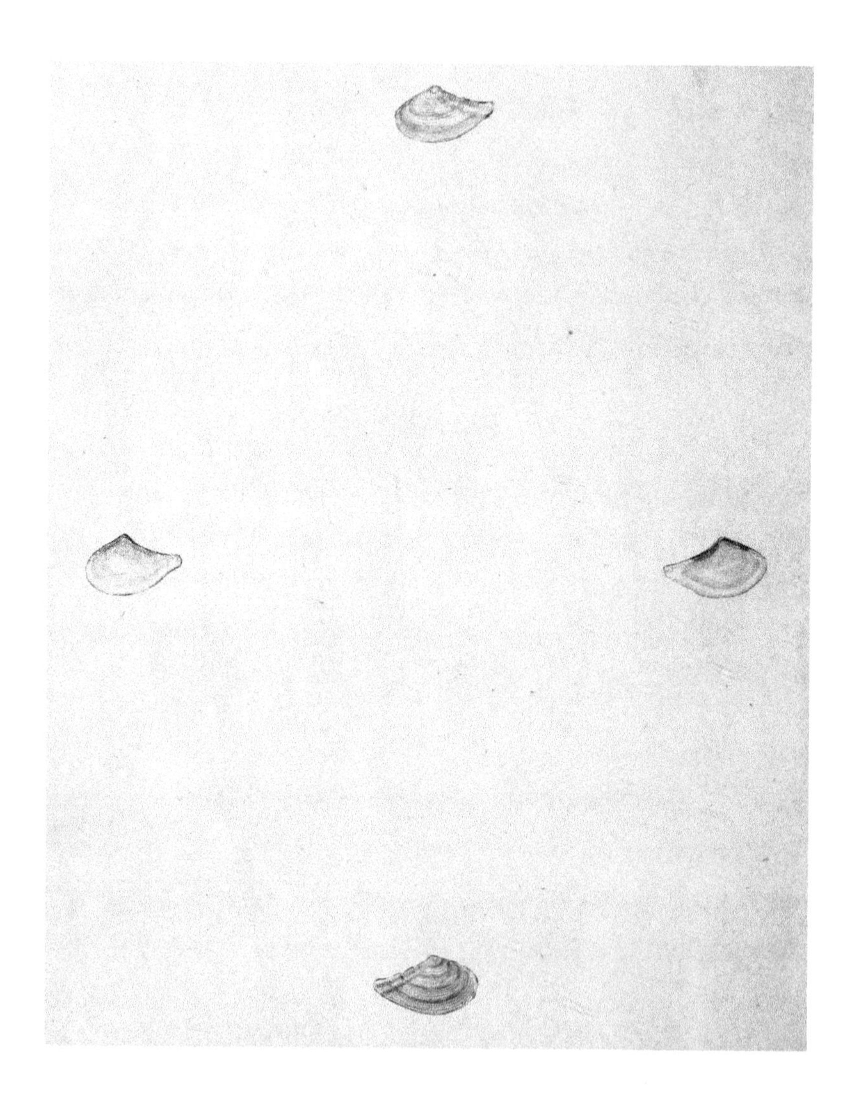

PLATE LXXVIII.

ARCA CAUDATA.

TAILED ARK SHELL.

GENERIC CHARACTER.

Animal Tethys. Shell bivalve equivalve. Teeth of the hinge numerous, inserted between each other.

SPECIFIC CHARACTER.

Oblong oval, one end rotundated, the other produced or lengthened out, angulated, and truncated at the end.

ARCA CAUDATA: testa oblongo ovali anterius rotundata posterius elongata angulata, apice subtruncata.

Very rare, and not hitherto described as a British species. Found on the Kentish coast.

79

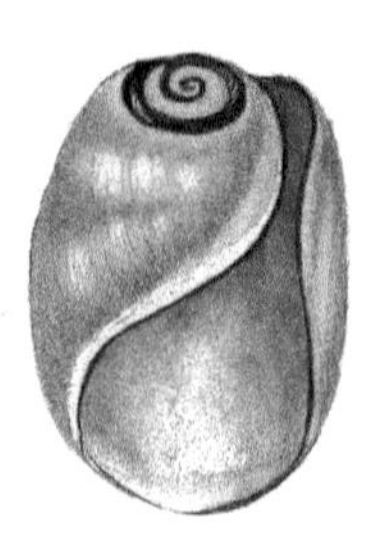

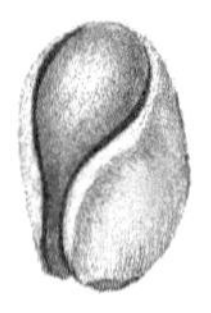

PLATE LXXIX.

BULLA RESILIENS.

ELASTIC BULLA.

GENERIC CHARACTER.

Animal Limax. Shell rather convoluted at one end, sub-oval. Aperture oblong.

Shell oval, pellucid, elastic. Spire somewhat depressed and canaliculated, or grooved along the margin.

BULLA RESILIENS: ovalis, pellucida, vi resiliendi prædita, spira, subdepressa anfractibus canaliculatus.

This interesting species of Bulla, which has lately been found in Devonshire, and considered as a new discovery, was first introduced to the notice of English Conchologists by the Rev. Mr. Cordiner. He discovered them some years ago on the shores of Bamff, and sent them, with several others, disposed in a grotto work, as a present to the late Duchess of Portland. Since that time they have been found at Lymington, in Hampshire, by

Mr. Keate; and lastly, during the summer months of 1800, was taken in a moat near Portsmouth, by J. Laskey, Esq. of Crediton, who favoured us with some particulars respecting the animal inhabiting it. In a young state, he says, it has the appearance of a winged insect, and sports in its watery element with all the liveliness of a butterfly, and formed a pleasing object when kept alive in a glass of sea water. It seems to prefer little pools, or still waters within reach of the tide, to more exposed situations.

In general the specimens that have been found at Portsmouth are very small, the shell from which the upper figure is copied far exceeding the others in size. This species, though very thin and brittle, is yet so elastic as to bear much compression without injury, and in this respect differs from every other British species of Bulla already known. Amongst the foreign kinds are several elastic kinds; and this very species is found of a much larger size in the Mediterranean Sea.—Independent of its elasticity, the convoluted apex is a material character of this shell, considered as a British species.

80

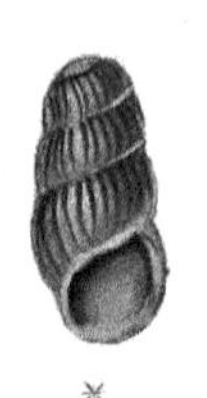

PLATE LXXX.

TURBO MUSCORUM.

CYLINDRIC, OR MOSS WREATH SHELL.

GENERIC CHARACTER.

Animal Limax. Univalve, spiral or of a taper form. Aperture somewhat compressed, orbicular, entire.

SPECIFIC CHARACTER.

Cylindric, pellucid, six spires, separated by a strong furrow, obtuse at the tip. Mouth oval.

TURBO MUSCORUM: testa ovata obtusa pellucida: anfractibus senis secundis, aperture edentula. *Gmel. Linn. Syst. Nat. p.* 3611. *sp.* 94.

Cochlea testa subpellucida, spiris sex dextrorsis, subcylindracea obtusa, *Linn. Faun. Suec.* I. *p.* 372. *No.* 1301. 2. *No.* 2173.

Turbo minimus mucrone obtuso, sive vere cylindraceus. *Cylindraceus, tab.* 5. *fig.* 16.

Buccinum exiguum subflavum, mucrone obtuso, sive cylindraceum. *List. H. An. Angl. p.* 121. *tit.* 6. *tab.* 2. *fig.* 6.

Buccinulum minimum ovale. *Petiv. Gaz. tab.* 35. *fig.* 6.
Morton, Northampt. p. 415.

Turbo Muscorum. *Pen. Br. Zool. No.* 118. *tab.* 82. *fig.* 118?

PLATE LXXX.

Linnæus and Da Costa have described this species with six spires; Pennant mentions only four; and we have remarked, that those with four spires are more numerous than the others.

It is a small shell, rarely exceeding the size of the smallest figures in the annexed plate; is very transparent, smooth and glossy, but under the magnifier exhibits many longitudinal streaks.

This shell inhabits mosses on old walls, thatches, trees, &c. It has been found by Da Costa in Middlesex and Surry; by Petiver on the sandy banks of the Thames, at Kingston, in the latter county. Dr. Lister, in plenty at Estrope, in Lincolnshire. Morton, in great plenty in the ground near Morsley Wood, in Northamptonshire; and received also by Da Costa, from Leeswood, in Flintshire.

81

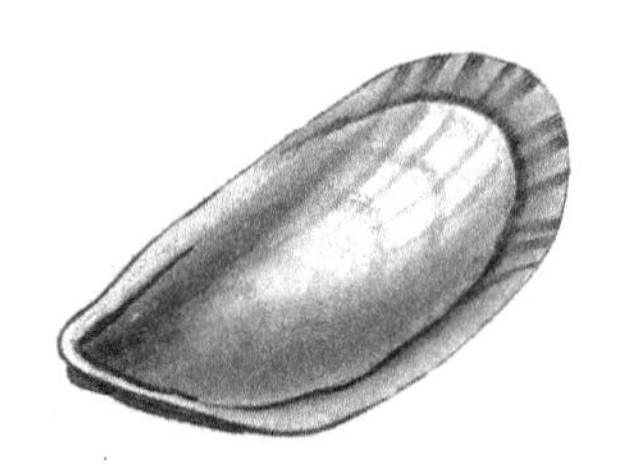

PLATE LXXXI.

MYTILUS PELLUCIDUS.

PELLUCID MUSCLE.

GENERIC CHARACTER.

The hinge toothless and consists of a longitudinal furrow.

SPECIFIC CHARACTER

AND

SYNONYMS.

Oblong, very pellucid, rayed longitudinally with purple.

MYTILUS PELLUCIDUS: oblonga pellucida longitudinaliter violaceo-radiata.

MYTILUS PELLUCIDUS. *Penn. Brit. Zool.* 4. *p.* 112. *sp.* 75.

This is one of the new species of Mytilus discovered by Pennant on the coast of Anglesea, where he says, it is "found sometimes in oyster-beds, and sometimes in trawling over slutchy bottoms." We dredged up a specimen of it in the straits of Menai, but it was rather less of an oblong form than that described and figured by Pennant; and another similar to it was also found on the Flintshire shores:— both Pennant's specimen and ours are figured in the annexed plate,

PLATE LXXXI.

We have lately received a very analogous species, if not a mere variety of it from Portsmouth; but those were evidently of foreign growth, having been gathered from the bottom of the William Tell prize ship, soon after its arrival from Malta.

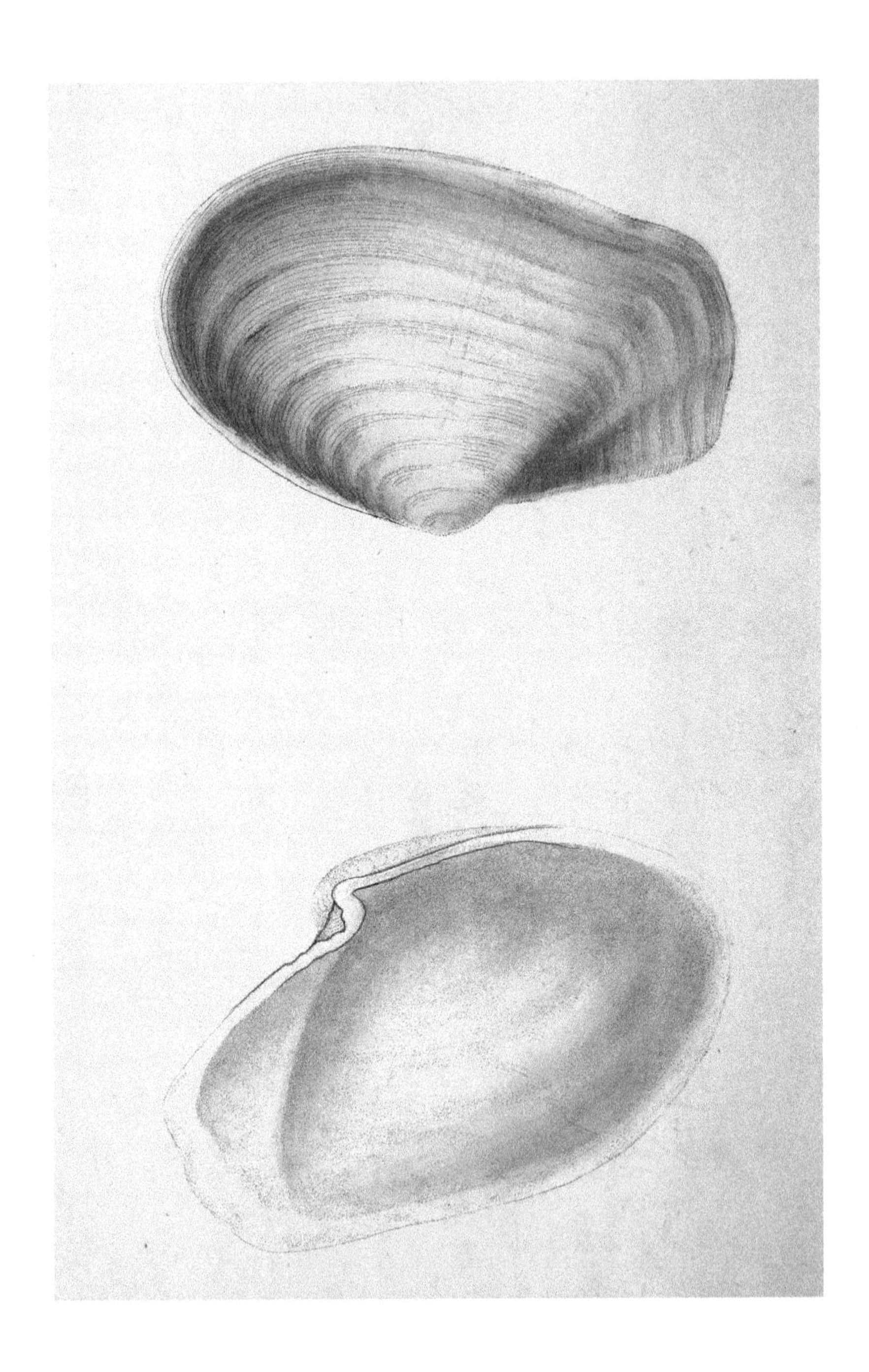

PLATE LXXXII.

MYA DECLIVIS.

SLOPING MYA, OR GAPER.

GENERIC CHARACTER.

Animal an Ascidia. Shell bivalve gaping at one end. The hinge for the most part furnished with a thick, strong, and broad tooth, not inserted into the opposite valve.

SPECIFIC CHARACTER

AND

SYNONYMS.

Shell somewhat oval, posterior end obliquely angulated. Tooth of the hinge thick and scarcely prominent.

MYA DECLIVIS: testa subovali, postice oblique subangulata. Cardinis dente crasso vix prominente.

MYA DECLIVIS with a brittle, half transparent shell, with a hinge slightly prominent; less gaping than the truncata; near the open end sloping downwards. *Penn. Br. Zool. Vol. 4. p. 79. sp. 15.*

C 2

PLATE LXXXII.

Pennant informs us that this species is frequent about the Hebrides; and that the fish is eaten by the gentry. We cannot question his authority in this respect, but must observe, it is uncommonly rare in cabinets of British Shells, and has not even been noticed, we believe, by any other Author.—Pennant has not figured it.

PLATE LXXXIII.

VENUS GRANULATA.

SPECKLED VENUS.

GENERIC CHARACTER.

Bivalve. Hinge furnished with three teeth; two near each other, and the third divergent from the beaks.

SPECIFIC CHARACTER

AND

SYNONYMS.

Shell rotund, sulcated longitudinally and decussated with transverse striæ; margins crenulated. Outside whitish, variegated with livid and purple spots. Inside violet.

VENUS GRANULATA: testa rotundata decuſſatim striata anterius et margine crenulato violacea. *Gmel. Linn. Syst. Nat. p.* 3277. *sp.* 33.

Venus marica. *Born. Mus. Cæs. vind. test. t.* 4. *f.* 5. 6.

Born has figured and described this shell as Venus Marica, and to distinguish it from a Linnæan species of the same name some conchologists have denominated it Venus Marica spuria. It is the Venus granulata of Gmelin, who refers to Born's figure in the synonyms.

Gmelin describes another shell under the name of Venus Violacea; which nearly agrees with V. granulata, V. VIOLACEA: testa intus violacea: striis perpendicularibus squamosis, margine denticulato. *Gmel. Syst. Nat. p.* 3288. *sp.* 94. This shell is figured in *Lister's Conch. t.* 338. *f.* 175. and is destitute of those external marks and specklings we have invariably observed on specimens of Venus granulata.

V. granulata is very rare on our coast. The smallest shell in the annexed plate was found in Cornwall. The large specimen is probably an old shell of this species.

84

PLATE LXXXIV.

HELIX POMATIA.

ITALIAN OR EXOTIC SNAIL.

GENERIC CHARACTER.

Aperture of the mouth contracted and lunulated.

SPECIFIC CHARACTER

AND

SYNONYMS.

Shell globose, with five spires, and umbilicated; whitish fasciated with brown. Mouth rather roundish.

HELIX POMATIA: testa umbilicata subovata obtusa decolore, apertura subrotundo-lunata. *Gmel. Linn. Syst. Nat. p.* 3627. *sp.* 47. *Fn. Suec.* 1283.

Cochlea magna cinereo rufescens, fasciata, leviter umbilicata. POMATIA. *Da Costa. Br. Conch. p.* 67. *sp.* 38. *tab.* 4. *fig.* 14. 14.

Cochlea cinerea, maxima, edulis, cujus os operculo crasso velut Gypseo per hyemem clauditur. Pomatia. *Gesn. de Aquat. pp.* 644. 255.

Cochlea cinereo rufescens, fasciata, leviter umbilicata. Pomatia Gesneri. *List. H. Conch. tab.* 48. *fig.* 46.

C 4

PLATE LXXXIV.

Cochlea pomatia edulis Gesneri. *List. Exercit. Anat.* 1. *p.* 162. *tab.* 1.

Cochlea alba major cum suo operculo. *Merret Pin. p.* 207.
Morton Northampt. p. 415.
Rutty Dublin. p. 379.

Helix Pomatia, Exotic. *Penn. Br. Zool. No.* 128. *tab.* 84, *fig.* 128.

Pomatia. *Argenville Conch.* I. *tab.* 32. *fig.* 1. *p.* 383. II. *p.* 338. *tab.*,28. *fig.* 1. *p.* 81. *tab.* 9. *fig.* 4.

Helix testa imperforata globosa rufescente, fasciis obsoletis. *Mul. Zool. dan. prodr.* 2901. *Hist. verm.* 2. *p.* 43. *n.* 243.

Cochlea testa ovata quinque spirarum, pomatia dicta. *Linn. Fn. Suec.* 1. *p.* 369. *No.* 1293. II. *No.* 2183.
Martin berl. Mag. 2. *p.* 530. *tab.* 1. *fig.* 1. *et.* 3. *tab.* 2. *fig.* 13.
Schroet. Erdconch. p. 145. *n.* 14. 15. *tab.* 1. *fig.* 10.
Knorr Vergn. 1. *tab.* 21. *fig.* 32.

Pennant has named this species of Helix with some propriety the Exotic Snail, for, though it is found at this time in vast abundance in several parts of the country, it is not an indigenous kind. By whom it was first introduced is uncertain; Pennant mentions Sir Kenelm Digby, and Da Costa speaks of Charles Howard, Esq. of the Arundel family. Its history, as related by Da Costa, is so very interesting, that we shall give it in the words of its author:—

" It is the largest species of land snail in England, and is found in hedges and woods. It closes its shell carefully against winter, with a white thick cover or operculum, dull and like plaister, and in the closed state it remains till the beginning of April, or warm weather, at which time it loosens the border of the cover, and the animal creeps out of the shell for its necessary occasions. Dr. Lister informs us he kept one in his bosom about the beginning of March, when the animal, feeling the warmth, in a few hours disengaged its cover, and crept out.

" The animal being large, fleshy, and not of an unpleasant taste, has been used for food in ancient times: it was a favourite dish with the Romans, who had their cochlearia, or snail stews, wherein they bred and fattened them. Pliny tells us, that the first inventor of this luxury was a Fulvius Harpinus, a little before the civil wars between Cæsar and Pompey. Varro has handed down to us a description of the stews, and manner of making them: He says, open places were chose, surrounded by water, that the snails might not abandon them, and care was taken that the places were not much exposed to the sun, or to the dews. The artificial stews were generally made under rocks or eminences, whose bottoms were watered by lakes or rivers; and if a natural dew or moisture was not found, they formed an artificial one, by bringing a pipe to it bored full of holes, like a watering pot, by which the place was continually sprinkled or moistened. The snails required little attention or food, for as they crawled they found it on the floor or area, and on the walls or sides, if not hindered by the surrounding water. They were fed with bran and sodden lees of wines, or like substances, and a few laurel leaves were thrown on it.

" Pliny tells us there were many sorts, as the Whitish from Umbria, the large sort from Dalmatia, and the African, &c. This particular kind seems to be that he mentions, l. 8. c. 39. They propagate very much, and their spawn is very minute.

" Varro is scarcely to be credited, when he says, some would grow so large, that their shells held ten quarts.

" They were also fed and fattened in large pots or pans, stuck full of holes to let in the air, and lined with bran and sodden lees, or vegetables.

" They are yet used as food in several parts of Europe, more especially during Lent, and are preserved in ftews or *escargotoires*, now a large place boarded in, and the floor covered with herbs, wherein they nestle and feed.

" In Italy, in many places, they are sold in the markets, and are called *Bavoli*, *Martinacci* and *Gallinelle*; in many provinces of France, as Narbonne, Franche Comté, &c. and even in Paris. They boil them, says Lister, in river water, and seasoning them with salt, pepper, and oil, make a hearty repast.

" This is not indigenous, or originally a native of these kingdoms, but a naturalized species, that has throve so well, as now to be found in very great quantities. It was first imported to us from Italy about the middle of last century, by a *scavoir vivre*, or epicure, as an article of food. Mr. Aubrey informs us, it was a Charles Howard, Esq. of the Arundel family, who, on that account, scattered and dispersed those snails all over the downs, and in the woods, &c. at Albury, an ancient seat of that noble family, near Ashted, Boxhill, Dorking, and Ebbisham, or Epsom, in Surrey, where they have thriven so much that all that part of the county, even to the confines of Sussex, abounds with them ; insomuch that they are a nuisance, and far surpass in number the common, or any other species of English snails.

The Epicures, or *scavoir vivre*, of those days, followed this luxurious folly, and the snails were scattered or dispersed throughout the kingdom, but not with equal success; neither have records transmitted to posterity the fame of those worthies equal to the Roman Fulvius Harpinus, except of two, the one Sir Kenelm Digby, who dispersed them about Gothurst the seat of that family (now of the Wrights) near Newport Pagnel, in Buckinghamshire, where probably they did not thrive much, as they were not frequent thereabout : the other worthy was a lord Hatton, recorded by Mr. Morton, who scattered them in the coppices at his seat at Kirby, in Northamptonshire, where they did not succeed.

" Dr. Lister found them about Puckeridge and *Ware*, in Hertfordshire ; and observes, they are abundant in the Southern parts, but are not found in the northern parts of this island.

" In Surry, as before mentioned, they abound ; in several other counties they are not uncommon, as in Oxfordshire, especially about Woodstock and Bladen ; in Gloucefterfhire, in Chedworth parish, and about Frog Mill, in Dorsetshire, &c. but I have never heard that they are yet met with in any of the northern counties."

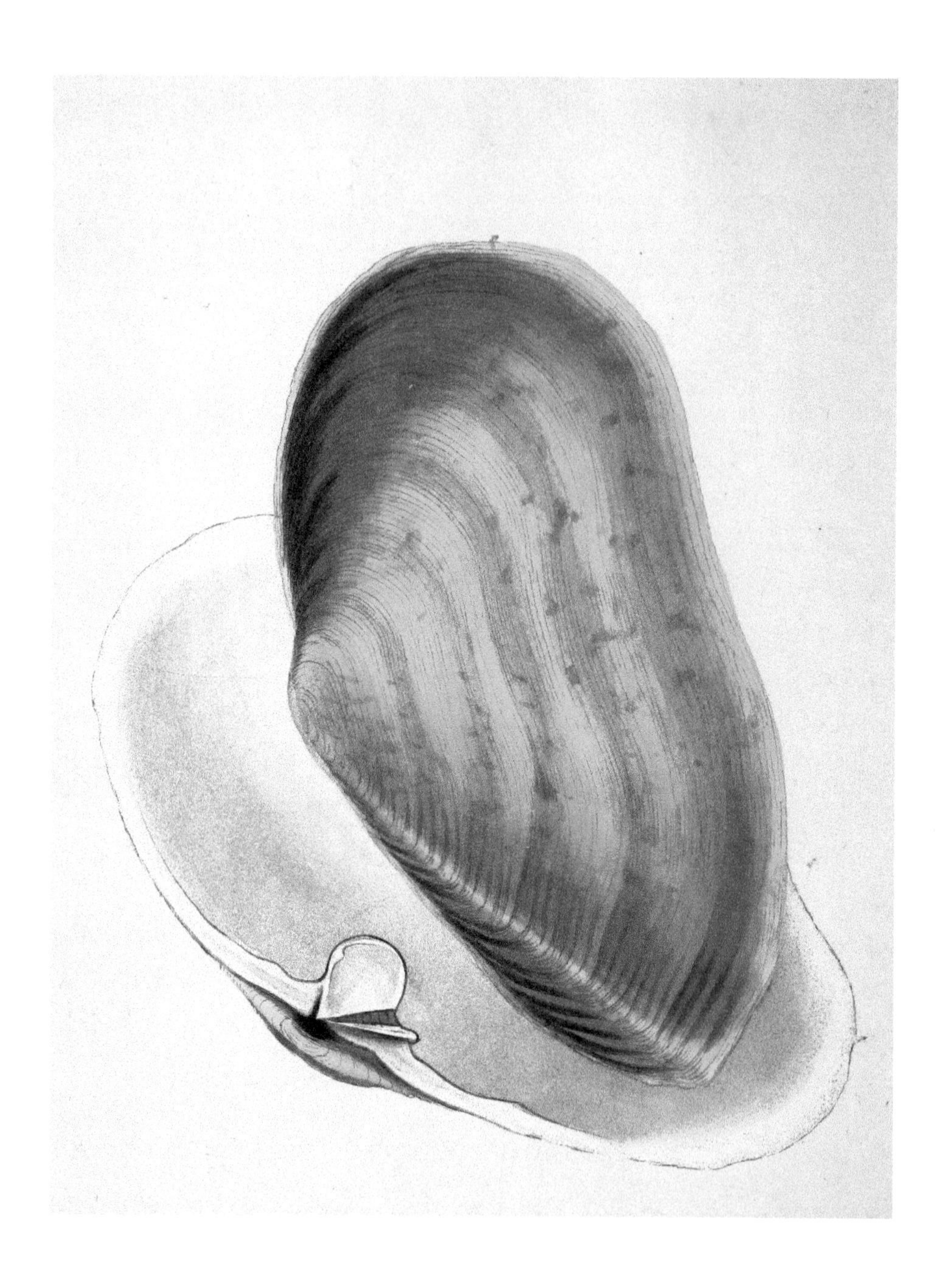

PLATE LXXXV.

MYA ARENARIA.

SAND GAPER.

GENERIC CHARACTER.

Animal an Ascidia. Shell bivalve, gaping at one end. The hinge for the most part furnished with a thick, strong, broad tooth, not inserted in the opposite valve.

SPECIFIC CHARACTER

AND

SYNONYMS.

Shell rather ovated, one end rounded, the other narrow and gaping. Hinge, in one valve a hollow cavity, near which a broad, erect, rounded tooth of the opposite valve is received.

MYA ARENARIA: testa ovata posterius rotundata, cardinis dente antrorsum porrecto rotundato denticuloque laterali. *Lin. Faun. Suec.* 2127.—*Gmel. Linn. Syst. p.* 3218. 303. *sp.* 2.

Mya Arenaria. Sand. *Penn. Br. Zool. p.* 79. *T.* 42. 16.

Chamæ media ovata fusca. Arenaria. *Da Cofta. Br. Conch. p.* 232. *sp.* 56.

MYA ARENARIA. *Bast. opusc. subs.* 2. *p.* 69. *t.* 7. *fig.* 1-3.

PLATE LXXXV.

This species is similar in its external appearance to the Mactra Lutraria; yet it may be immediately distinguished from that shell by the singular structure of the hinge. The large, erect, plate-like tooth common to the Mya genus, is particularly characteristic in this species.

Da Costa received it from the Isle of Wight, near Newport, and from Bigbury-Bay, near Faversham; but observes, it is not a common shell.

86

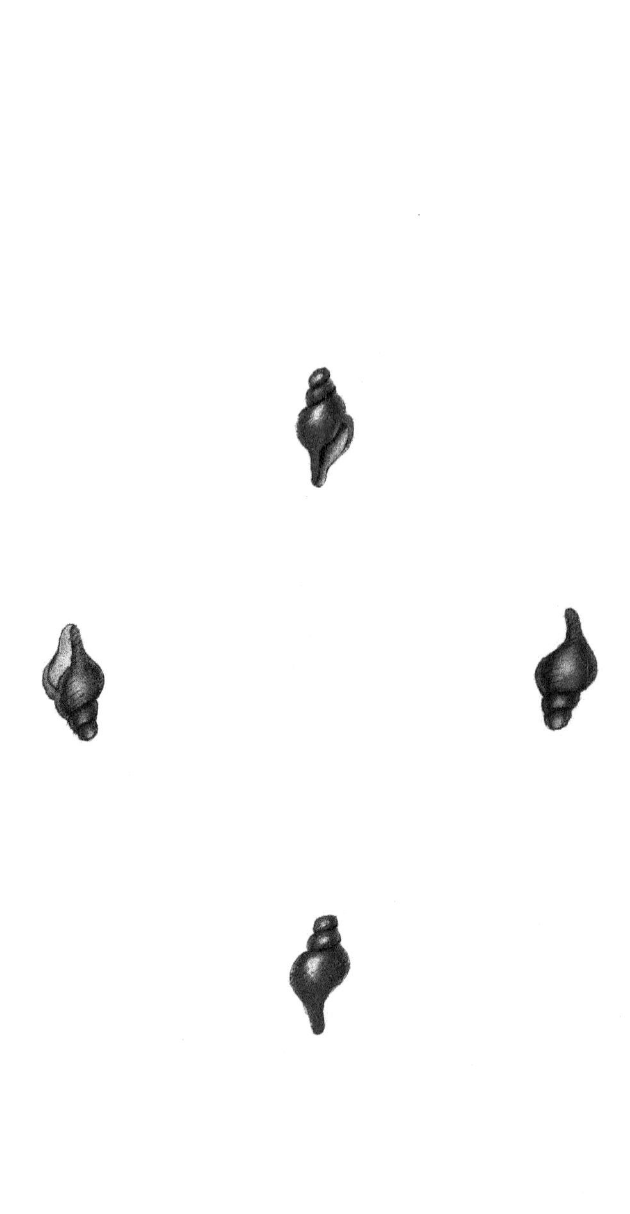

PLATE LXXXVI.

MURFX DECOLLATUS.

SHORTENED MUREX, OR ROCK SHELL.

GENERIC CHARACTER.

Spiral, rough, the aperture ending in a strait, and somewhat produced gutter or canaliculation.

SPECIFIC CHARACTER.

MUREX DECOLLATUS: testa ventricosa lævi, patulo-subcaudata, spira in capitulum desinente.

Somewhat ventricose, smooth, mouth produced. Spire terminated in a capitulum or knob.

MUREX DECOLLATUS. *Penn. Br. Zool. T.* 4. *p.* 125. *sp.* 102.

Pennant offers his Murex Decollatus as a species with doubts. It has certainly the appearance of a shell much mutilated, or of extraordinary growth; but as all the specimens we have examined exhibit the same appearance, we have ventured to assign it a new character, and rank it as a distinct species.

It is a rare shell on the British shores, said to have been found on those of Cornwall and Devonshire.

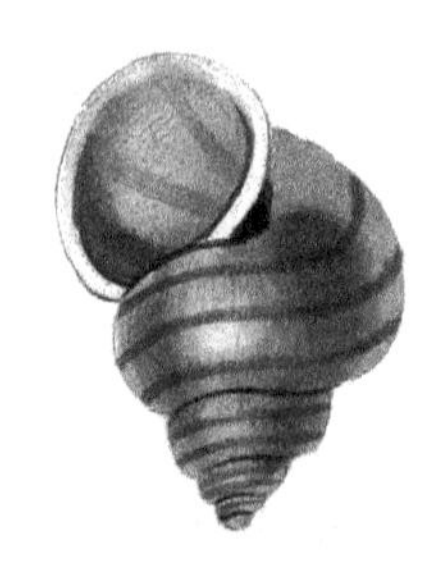

PLATE LXXXVII.

HELIX VIVIPARA.

VIVIPAROUS SNAIL.

GENERIC CHARACTER.

Aperture of the mouth contracted and lunulated.

SPECIFIC CHARACTER

AND

SYNONYMS.

Shell suboval, obtuse, spires ventricose or swelled, umbilicated. Olive, girdled with three brown lines.

HELIX VIVIPARA: testa imperforata subovata obtusa cornea: cingulis fuscatis, apertura suborbiculari *Fn. Su.* 2185.—*Gmel. Linn. Syst. Nat. p.* 3646. *sp.* 105.

Cochlea testa oblongiuscula obtusa anfractibus teretibus, lineis tribus lividis. *Fn. Suec. I. p.* 375. *No.* 1312.

Cochlea maxima fusca sive nigricans, fasciata. *List. H. An. Angl. p.* 133. *tit.* 18. *tab.* 2. *fig.* 18.

Cochlea fasciata ore ad amussim rotundo. *Phil. Trans. No.* 105. *fig.* 17.—Cochlea maxima viridescens fasciata vivipara. *List. Exercit. Anat.* 2. *p.* 17. *tab.* 2.

C. vivipara fasciata fluviatilis. *List. H. Conch. tab.* 126. *fig.* 26.—C. vivipara altera nostras testa tenuiori fluvii Cham. *Ib. Mant. tab.* 1055. *fig. C.*

PLATE LXXXVII.

Helix vivipara, viviparous. *Penn. Br. Zool. No.* 132. *tab.* 84. *fig.* 132.

Cochlea fusco viridescens trifasciata. Vivipara. *Da Costa. Br. Conch. p.* 81. *sp.* 44.

This kind is found in abundance in all rivers and stagnant waters. The river kind seems to vary in some respects from the other; the shells are more opake, and the colours are brighter than in those which inhabit the stagnant water.

The animal has a head not unlike that of a Bull, from which circumstance the Swedes, according to Linnæus, call it the Bull-head, and some French authors, *limaçon à tete de bœuf*, for the same reason. It feeds on Duck Weed.

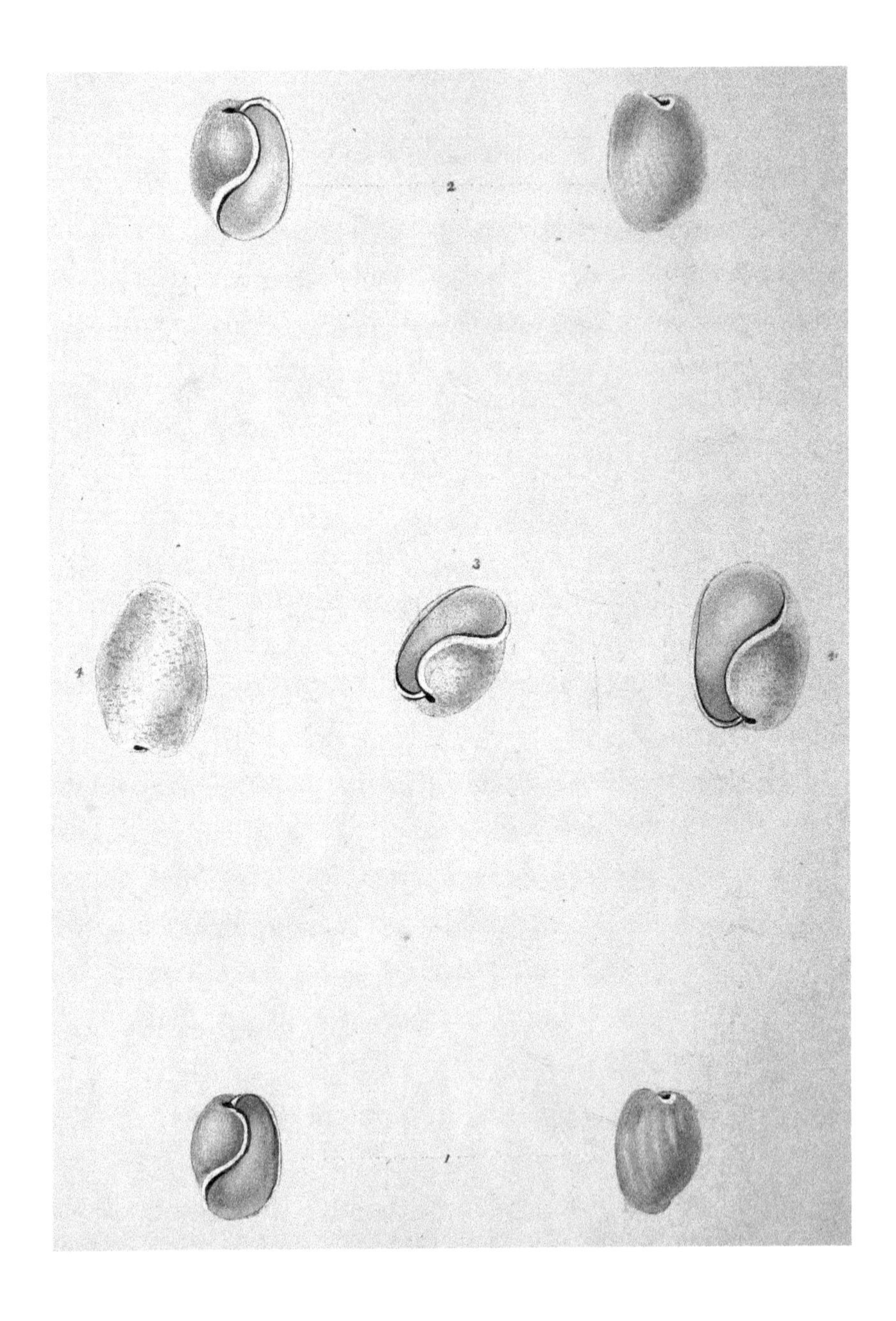

PLATE LXXXVIII.

BULLA HYDATIS.

PINNACE DIPPER, OR BUBBLE SHELL.

GENERIC CHARACTER.

Animal Limax. Shell rather convoluted at one end, sub-oval, Aperture oblong.

SPECIFIC CHARACTER AND SYNONYMS.

Oblong-oval, fragile, pellucid, finely striated longitudinally, base deeply umbilicated.

BULLA HYDATIS: testa rotundata pellucida longitudinaliter substriata: vertice umbilicato. *Linn. Syst. Nat. p.* 1183. *No.* 377.—*Gmel. Linn. Syst. Nat.* 3424. *sp.* 9.

Nux Marina umbilicata, minutissime per longitudinem striata, subrotunda, ore admodum patulo, tenius, fragilis candida. *Gualt.* 1. *Conch. tab.* 13. *fig. D. D.* *Chemn.* 9. *t.* 118. *f.* 1019.

Bulles d'eau blanches, papyracées. Tonnes à bouche entiere. *D'Avila Cab. p.* 207. *No.* 389.

Bulla Ovalis, fragilis et pellucida. Naviacula. *Da Costa Br. Conch. p.* 28. *sp.* 15.—*Tab.* 1. *fig.* 10.

PLATE LXXXVIII.

Da Costa observes, that all the shells he had seen of this species were fished up at, or near, Weymouth in Dorsetshire; and concludes, that it is rare in our seas, having never heard of it on any other British coast. We believe with Da Costa, it is local; though it probably inhabits other parts of our coasts.

We have been lately favoured with several shells of the Bulla genus from Portsmouth, which some Conchologists have thought a new species, and named Citrina; they do not, certainly, differ specifically from the shell in Da Costa's collection, which he calls Bulla Naviacula, (Hydatis of Linnæus) as will appear evident from the specimens figured in the annexed plate.

Fig. 1.—Bulla Naviacula *(Hydatis Linn.)*——Fig. 2, a specimen from Portsmouth of a paler colour than Da Costa's shell.——Fig. 3, 4. old shells found on the mud and clay of the shore.

89

PLATE LXXXIX.

MYA OVALIS.

OVAL GAPER.

GENERIC CHARACTER.

Animal ascidia. Shell bivalve, gaping at one end. The hinge for the most part furnished with a thick, strong, broad tooth, not inserted in the opposite valve.

SPECIFIC CHARACTER
AND
SYNONYMS.

Shell rather an oblong oval, with a large longitudinal crenulated tooth in one valve, and two in the other.

MYA OVALIS: testa oblongo-ovali cardinis dente primario crenulato longitudinali: alterius duplicato.

Mususculus angustior, ex flavo viri descens, validus, umbonibus acutis, valvarum cardinibus velut pinnis donatis, sinuosis. *List. Angl. t.* 2. *f.* 30.

Long thick horse Muscle. *Petiv. Gaz. tab.* 93. *fig.* 9

Mya pictorum. *Penn. Br. Zool.* 43. *fig.* 17.

E 2

PLATE LXXXIX.

Mya minor ex flavo viridescens. PICTORUM *Da Costa. Br. Conch. p.* 228. *tab.* 14. *fig.* 4. 4.

Pennant and Da Costa have erroneously given this as the *Mya pictorum* of Linnæus, from which it differs in several respects. The Mya pictorum is much more ovate, or egg-shaped, as Linnæus expresses it, and thinner than the present shell, which is of a lengthened, or rather oblong form, and remarkably thick, though semi-transparent. In adopting a new specific name that of *oblonga* would have been preferred, had it not been pre-engaged by *Gmelin* to a totally distinct species.

This species is common in rivers and fresh waters, and sometimes produce little pearls.

90

PLATE XC.

TURBO LACTEUS.

SMALL TURBO.

GENERIC CHARACTER.

Animal Limax. Shell univalve, spiral, or of a taper form. Aperture somewhat compressed, orbicular, entire.

SPECIFIC CHARACTER AND SYNONYMS.

Shell taper, with many longitudinal, elevated striæ, or ridges.

TURBO LACTEUS: testa turrita: striis longitudinalibus elevatis confertis. *Linn. Syst. Nat. p.* 1238. *No.* 634.

Turbo parvus interdum lacteus, interdum violaceus aut fuscus, costis longitudinalibus confertus. Parvus. *Da Costa. Br. Conch. p.* 104. *sp.* 61.

Found on the coasts of Cornwall, Devonshire, and Guernsey.—This is a minute and scarce British species; the smallest figures in the annexed plate denote the natural size.

PLATE XC.

Some specimens are pure white, others beautifully tinged with purple; and the most perfect white and brown. The mouth is round, surrounded on the outside by a thick prominent border. It has no umbilicus. The shell consists of five spires, gradually tapering to an acute point; and separated by a depression. The longitudinal ribs are thick and prominent.

91

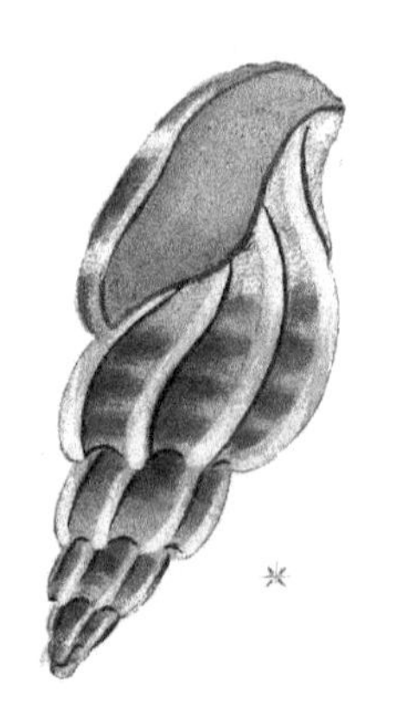

1 1

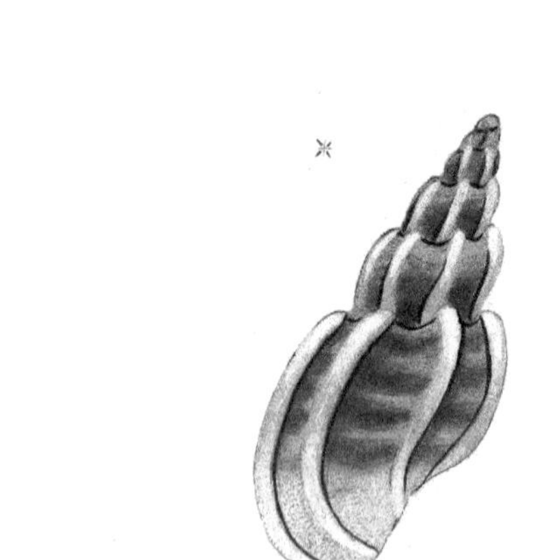

PLATE XCI.

MUREX COSTATUS

RIBBED MUREX.

GENERIC CHARACTER.

Spiral, rough. The aperture ending in a strait and somewhat produced gutter or canaliculation.

SPECIFIC CHARACTER

AND

SYNONYMS.

Oblong, spires six, tapering, with eight longitudinal ribs.

Buccinum canaliculatum parvum, anfractibus costis longitudinalibus distinctis. Costatum. *Da Costa. Tab.* 8. *fig.* 4.

MUREX COSTATUS. Ribbed. *Penn. Br. Zool. No.* 100. *tab.* 79. *fig.* 1. 4.

This elegant little shell was first discovered by Mr. Pennant, on the coast of Anglesea, and described under the name of *Murex Costatus.* In retaining this name it will be proper to observe, that *Gmelin*, in his edition of the Systema Naturæ, has another shell

E 4

under the same name, a ribbed and cancellated species found in a fossil state, in *Champagne*, altogether distinct from this shell.

Da Costa received this species from the coasts of Cornwall and Devonshire. Pennant says it inhabits Norway. The smallest figures denote the natural size.

92

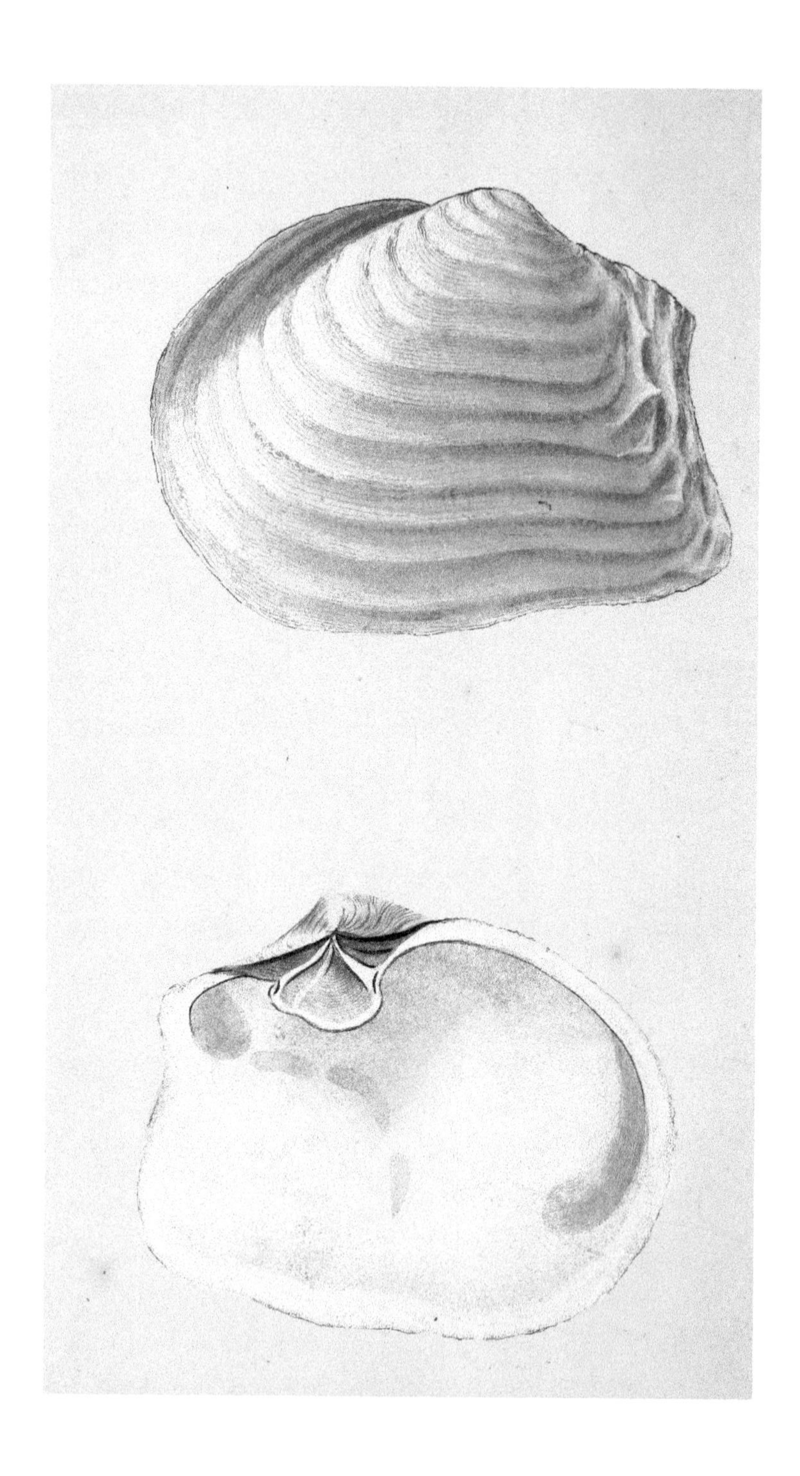

PLATE XCII.

MYA TRUNCTATA.

TRUNCATED GAPER.

GENERIC CHARACTER.

Animal an ascidia. Shell bivalve, gaping at one end. The hinge for the most part furnished with a thick, strong, broad tooth, not inserted into the opposite valve.

SPECIFIC CHARACTER

AND

SYNONYMS.

Shell roundish, one end trunctated or abrupt. Tooth projecting and obtuse.

MYA TRUNCTATA: testa ovata posterius truncata, cardinis dente antrorsum porrecto obtussissimo. *Linn. à. Gmel. Syst. Nat. T. I. fig.* 6. *p.* 3217.

Concha lævis, altera tantum parte clusilis, apophysi admodum prominente lataque prædita. *List. H. An. Angl. p.* 191. *tit.* 36. *tab.* 5. *fig.* 36.

PLATE XCII.

Mya truncata, abrupt. *Penn. Br. Zool.* 4. 14. *tab.* 41. *fig.* 14.
Chama subrotunda fusca rugosa, exaltera parte trunctata. TRUNC-
TATA. *Da Costa. Br. Conch. p.* 233. *sp.* 57.

Common on many of the British shores.

93

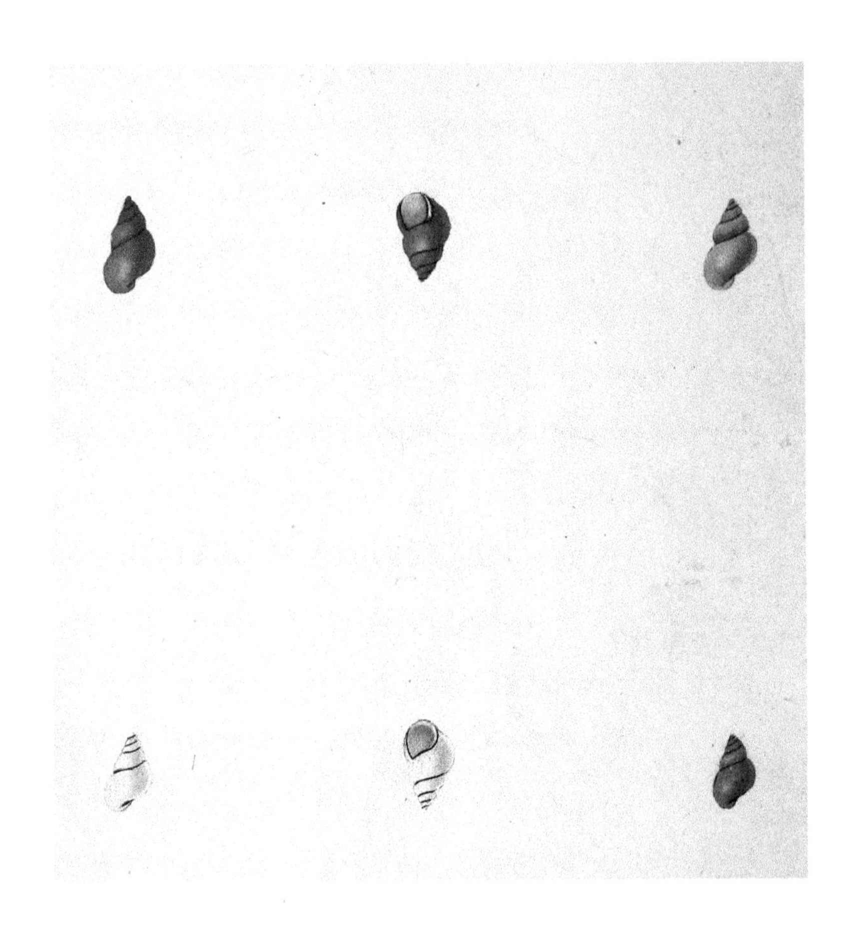

PLATE XCIII.

HELIX TENTACULATA.

KERNEL, OR OLIVE WATER SNAIL.

GENERIC CHARACTER.

Aperture of the mouth contracted and lunulated.

SPECIFIC CHARACTER

AND

SYNONYMS.

Shell without umbilicus, sub-conic, five spires. Aperture rather oval.

HELIX TENTACULATA: testa imperforata ovata obtusa impura, aperture subovata. *Linn. Syst. Nat. p.* 1249. *n.* 707.

Cochlea parva pellucida, operculo testaceo cochleatoque clausa. *List. H. Conch. tab.* 132. *fig.* 32.

Cochleola oblonga fluviatilis, common small river snail. *Petiv. Gaz. tab.* 18. *fig.* 8.—Small fresh water turbo with five wreaths. *Wallis Northumb. p.* 370.

Turbo imperforatus parvus subrufus, lævis, quinque spirarum. Nucleus. *Da Costa. Br. Conch. p.* 91. *sp.* 50.

PLATE XCIII.

Helix tentaculata. *Penn. Brit. Zool.* 4. *No.* 140. *tab.* 86. *fig.* 140.

Inhabits most rivers and stagnant waters.

1

1

PLATE XCIV.

STROMBUS COSTATUS.

RIBBED STROMBUS.

GENERIC CHARACTER.

Animal a slug. Shell univalve, spiral. Aperture dilated, lip expanding, produced into a groove.

SPECIFIC CHARACTER
AND
SYNONYMS.

Shell small, brown, taper, spires swelled, and wrought with prominent longitudinal ribs.

Strombiformis parvus fuscis, anfractibus costis elatis longitudinalibus insignitis.

COSTATUS. *Da Costa. Br. Conch. p.* 118. *sp.* 71.

Da Costa, who is the only author that describes this curious shell, says it is found on the coasts of Cornwall.

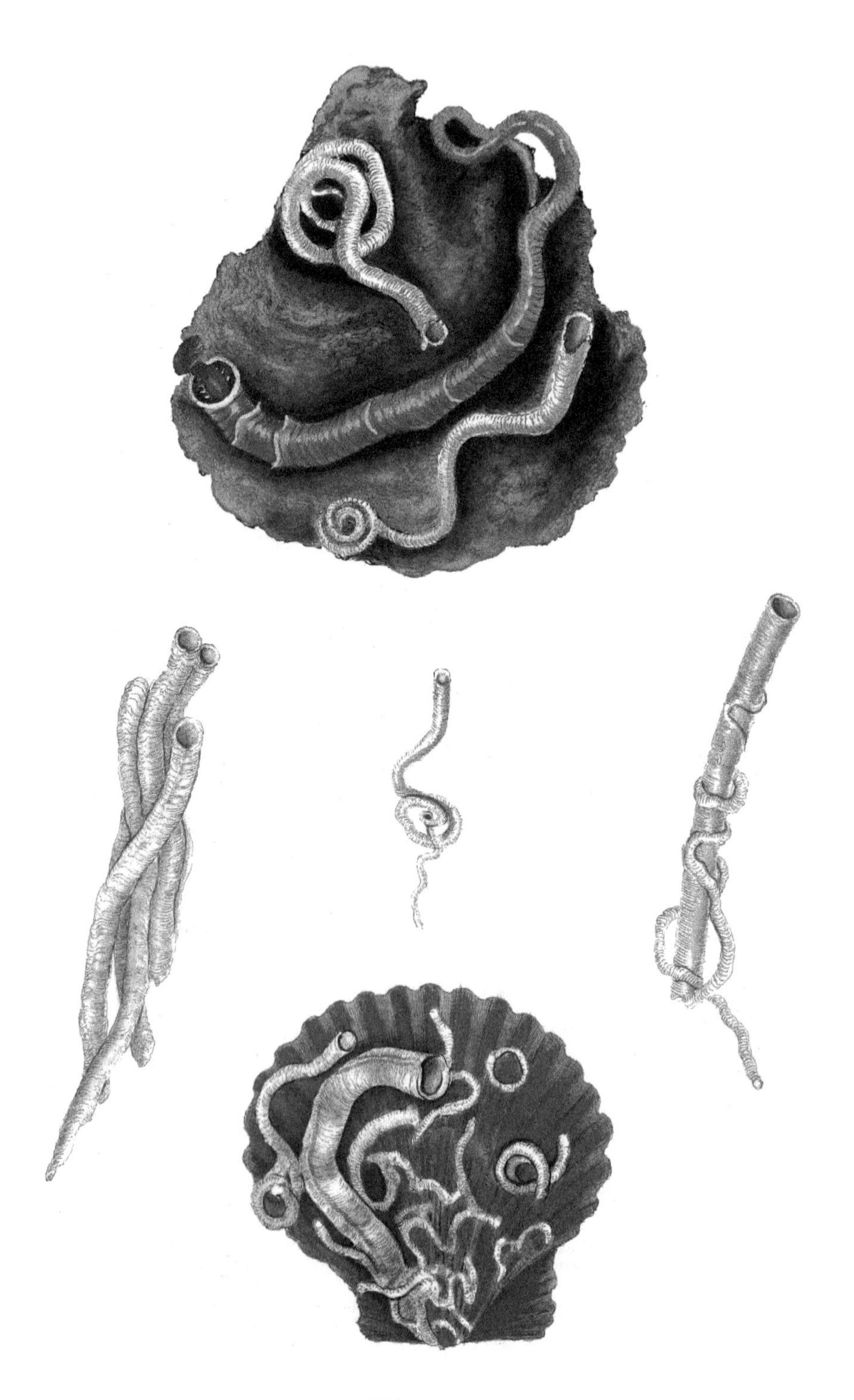

PLATE XCV.

SERPULA VERMICULARIS.

COMMON SERPULA.

GENERIC CHARACTER.

Animal a terebella, or whimble worm. Shell tubular, adheres to other bodies, as shells, stones, &c.

SPECIFIC CHARACTER
AND
SYNONYMS.

Shell round, cylindrical, or scarcely tapering, curved and wrinkled.

SERPULA VERMICULARIS: testa tereti subulata curvata rugosa. *Lin. Syst. Nat. a Gmel. T. I. fig.* 4. *p.* 3743. —Dentalium testa cylindracea inæquali flexuosa contorta. *Lin. Fn. Sv. I. p.* 380. *No.* 1328.

Tubuli in quibus vermes. Worm Shells. *Merret, Pin. p.* 194.

SERPULA VERMICULARIS. Worm. *Penn. Brit. Zool. No.* 157. *tab.* 91. *fig.* 159.

Serpulæ vermicularis, common. *Da Costa Br. Conch. p.* 18. *sp.* 9. —*Tab.* 2. *fig.* 5.

Those shells are extremely frequent on all the British coasts, either in groupes attached to stones, shells and marine exuviæ, or in single detached shells, assuming sometimes the appearance of a turbinated univalve.

The colour is in general white: an elegant variety, deeply tinged with red, as represented in the annexed plate, was dredged up at Brighton, and communicated by Mr. P. Munn, of Bond-street.

96

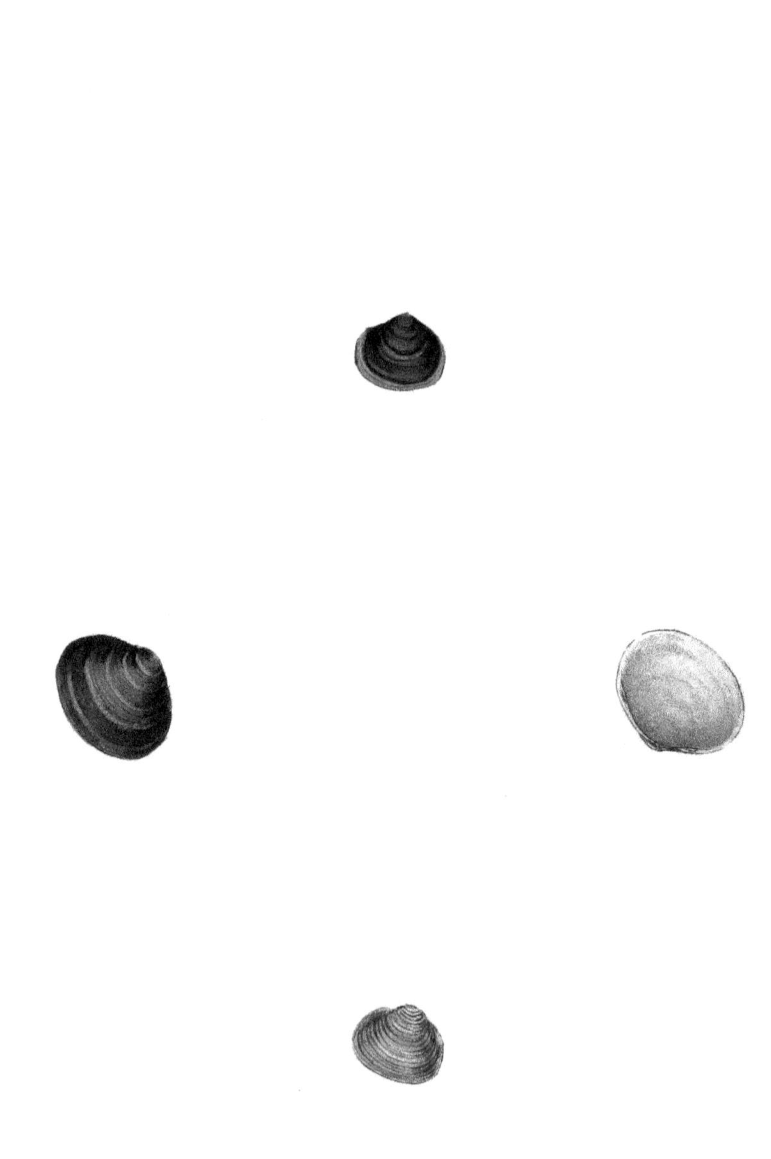

PLATE XCVI.

TELLINA CORNEA.

HORNY TELLEN.

GENERIC CHARACTER.

The hinge usually furnished with three teeth; shell generally sloping on one side.

SPECIFIC CHARACTER AND SYNONYMS.

Subglobose, glabrous, horn colour, with a transverse furrow.

TELLINA CORNEA: concha testa subglobosa glabra cornei coloris: sulco transversali. *Lin.*

Musculus exiguus, pisi magnitudine, rotundus subflavus, ipsis valvarum oris albidis. *List. H. An. Angl. p.* 150. *tit.* 31. *tab.* 2. *fig.* 31.

Pectunculus fluviatilis nostras nuciformis. *Petiv. Mus. p.* 86. *No.* 831.

Musculus fluviatilis, æquilaterus, lævis rotundus, pisiformis, ex rubro flavescens, ipsis valvarum oris albidis. *Gualt. I. Conch. tab.* 7. *fig.* C.

PLATE XCVI.

C. Parvum globosum viride-fuscum. Nux. *Da Costa Br. Conch.* 173.

Tellina Cornea. Horny. *Penn. Br. Zool. No.* 36. *tab.* 49. *fig.* 39.

Da Costa observes, that Linnæus has placed this shell very improperly in the Tellina genus, as it does not agree with his own definition of that genus, and remarks that its habit, shape, convexity, &c. brings it nearer to the Cardium than any other kind.—It still remains a Tellina in the last edition of the *Systema Naturæ* by Gmelin, and we are not disposed in the present instance to deviate from that authority.

This, and the Tellina rivalis described by Dr. Maton, in the Linnæan Transactions, are very analogous, though evidently two distinct species, as we have before noticed in our description of the latter, Plate 62.—Tellina Cornea, according to Geoffroy, is a viviparous animal, and is found in great plenty in most rivers and stagnant waters.

97

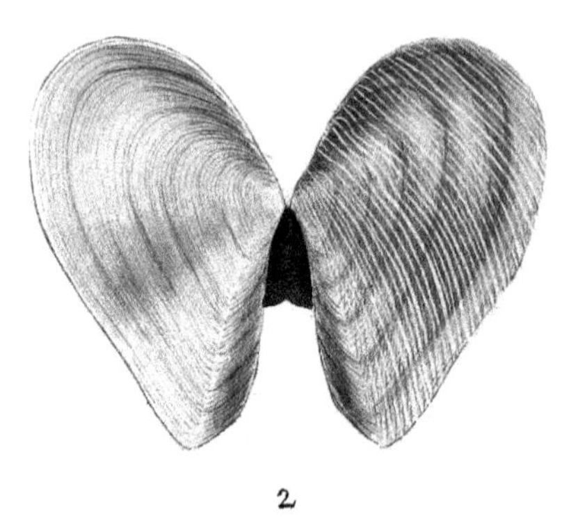

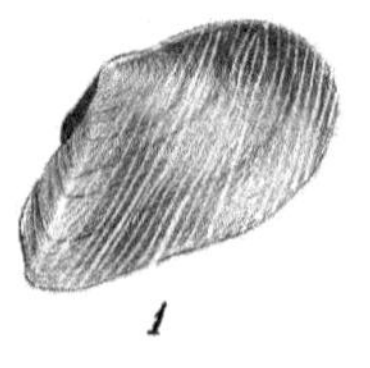

PLATE XCVII.

TELLINA FABULA.

SEMI-STRIATED TELLEN.

GENERIC CHARACTER.

The hinge usually furnished with three teeth. Shell generally sloping on one side.

SPECIFIC CHARACTER AND SYNONYMS.

Shell ovate, compressed, inflected, or rather produced at one end. One valve smooth, the other marked with numerous oblique reflected striæ.

TELLINA FABULA: testa ovata compressa inflexa anterius subrostrata: valva altera lævi, altera oblique substriata: striis reflexis.—*Gronov. Zooph. tab.* 13. *fig.* 9. *Gmel. Linn. Syst. Nat. T. I. p.* 6. *p.* 3239. *sp.* 61.

We discovered this very unusual species on the sands opposite to Caldy Island, about two miles beyond Tenby, Pembrokeshire. It is

PLATE XCVII.

noticed by Gronovius and Gmelin as a Norwegian and Mediterranean shell, and is said to have been found on the coast of Dorsetshire, by the late Dr. Pultney; but has never been before described as a British species.

The smallest figures represent the natural size.

38

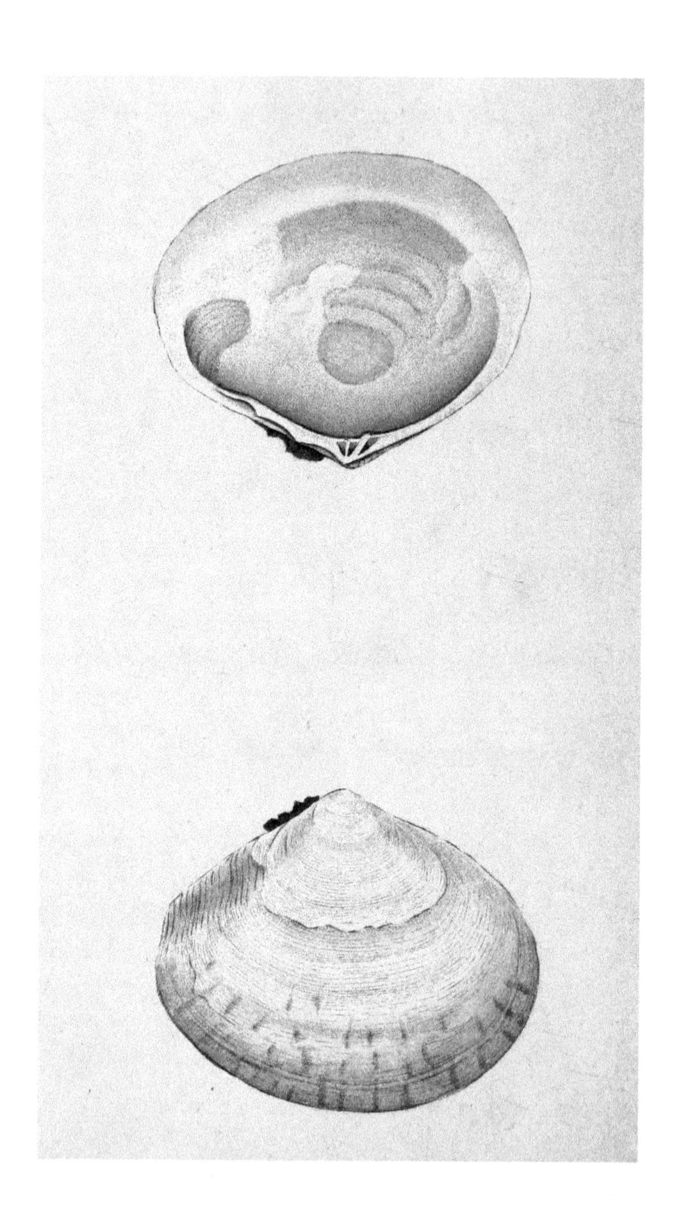

PLATE XCVIII.

TELLINA FAUSTA.

GENERIC CHARACTER.

The hinge usually furnished with three teeth. Shell generally sloping on one side.

SPECIFIC CHARACTER.

Somewhat triangular with many obsolete minute transverse striæ.

TELLINA FAUSTA: testa subtriangulari, striis transversis minutissimis obsoletis.

Tellina fausta. *Soland. Ms.—List. Conch. t.* 388. *f.* 235.

An extremely scarce British species, and not mentioned by either Penant, or Da Costa.

This shell is generally of a pale cream colour on the outside, and beautifully tinged with yellow within.

99

1

2

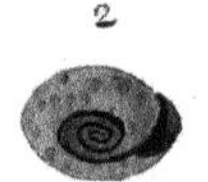

1

PLATE XCIX.

HELIX CONTORTA.

THICK RIVER CHEESE SHELL.

GENERIC CHARACTER.

Aperture of the mouth contracted and lunulated.

SPECIFIC CHARACTER

AND

SYNONYMS.

Shell thick, umbilicated, flattish. Aperture narrow and crescent-shaped.

HELIX CONTORTA: testa subumbilicata plana utrinque æquali: apertura lineari arcuata. *Gmel. Linn. Syst. Nat. p.* 3624.

Planorbis minima crassa Tiney many-circled, thick, river cheese shell. *Petiv. Gaz. tab.* 92. *fig.* 8.

Planorbis minima crassa, utrinque umbilicata, anfractibus subdepressis. Crassa. *Da Costa Br. Conch. p.* 66. *sp.* 37. *Tab.* 4. *fig.* 11.

G 2

PLATE XCIX.

This aquatic snail is rather scarce; it has been lately found in the Thames, near Greenwich. Petiver says his were found in the rivulets about Peterborough House, Westminster.

Da Costa mistook this for the Helix complanata of Linnæus: it is evidently the Helix contorta of that author, who very minutely describes it in the Fauna Suecica.

PLATE C.

SERPULA GRANULATA.

GENERIC CHARACTER.

Animal a terebella, or whimble worm. Shell tubular, adheres to other bodies, as shells, stones, &c.

SPECIFIC CHARACTER AND SYNONYMS.

Shell roundish, spiral, glomerate: three elevated ridges on the upper side.

SERPULA GRANULATA: testa tereti spirali glomerata; latere superiore sulcis tribus elevatis. *Gmel. Syst. T. I. p. 6. p.* 3741. *sp.* 9.

This singular species has not been before noticed as an English Shell. We found it intermixed with Lepas Intertexta on the shell of the common Lobster. Linnæus says it is found in the North Seas in large masses, adhering to stones, and shells.

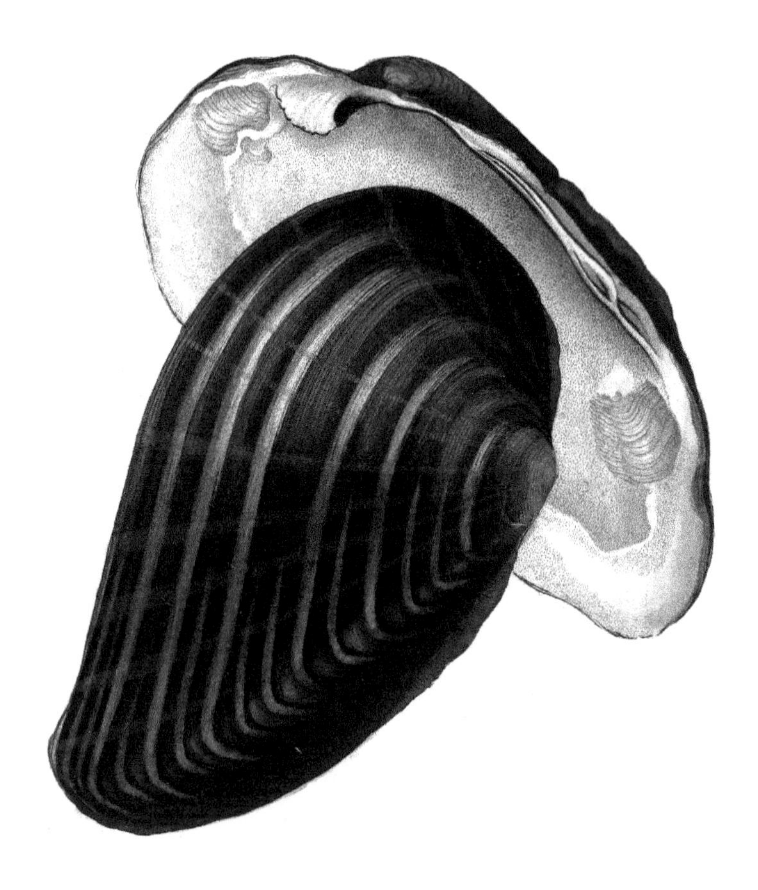

PLATE CI.

MYA DEPRESSA.

DEPRESSED MYA.

GENERIC CHARACTER.

Animal an ascidia. Shell bivalve, gaping at one end. The hinge for the most part furnished with a thick, strong, and broad tooth, not inserted into the opposite valve.

SPECIFIC CHARACTER.

Somewhat ovate, anterior part rather wedge-shaped and sloping: a slight depression across the middle; posterior part roundish, gaping. Teeth at the hinge crenulated.

MYA DEPRESSA: testa subovata, antice sub-cuneiformi declivi, medio depressa, postice rotundata hiante, cardinis dente crenulato.

After comparing the numerous kinds of fresh water Myæ found in different parts of the kingdom, the conchologist will perhaps be surprized at the small number we shall venture to admit as distinct species. The varieties of those Shells seem endless, and it may be

G 4

doubted whether they are not in general indebted to age, accident, or the peculiar qualities of the waters they inhabit, for those variations in general appearance that have been too frequently mistaken for characteristic differences of species.

As the Myæ will fall under consideration more fully hereafter, we shall for the present confine our remarks to the shell before us, and its very analogous kind, the Mya ovata of Dr. Solander.

This has been considered by some as a mere variety of ovata, and we confess our opinion is still wavering in assigning it a name and character as a new species. The Mya ovata has been lately found in the river Froome in Somersetshire, and likewise in the New River near London. What are usually deemed its varieties are numerous, but none of them can, we believe, be considered as distinct species, except the present, which is certainly the most remote of any, if it is really a variety of that species. The Mya ovata, in all its gradations, seems somewhat more ventricose and ovate in its contour, than this Shell; and though the variations of the latter are considerable, we have generally observed a slight depression, across the middle, which causes the narrowest end to be rather flattened throughout, and it is also rather more cuneiform or wedge-shaped at this end than Mya ovata: to this we might perhaps add, with some propriety, that the gaping beyond the hinge at the broadest end, is wider than in Mya ovata.

Whether this difference is actually sufficient to form a distinct specific character, and whether it is constant in other shells of this kind, still remains in some degree of uncertainty. Both this and the Mya ovata inhabit the same waters, for we have seen several specimens from the river Froome, where it is known the Mya ovata is also found; and as to colour, it is no criterion: both kinds are greenish, radiated with yellow, and are more or less vivid in different shells: they are seldom higher in colour than the specimen we have figured; some are more of an olive colour, and others are deeply tinged with brown.

PLATE CII.

TURBO FONTINALIS.

GENERIC CHARACTER.

Animal Limax. Univalve, spiral, or of a taper form. Aperture somewhat compressed, orbicular, entire.

SPECIFIC CHARACTER.

Shell umbilicated, subconic, wreaths ventricose, smooth.

Turbo Fontinalis: testa umbilicata subconica anfractibus ventricosis lævibus.

Not described by Pennant or Da Costa. Lives in clear fresh waters.

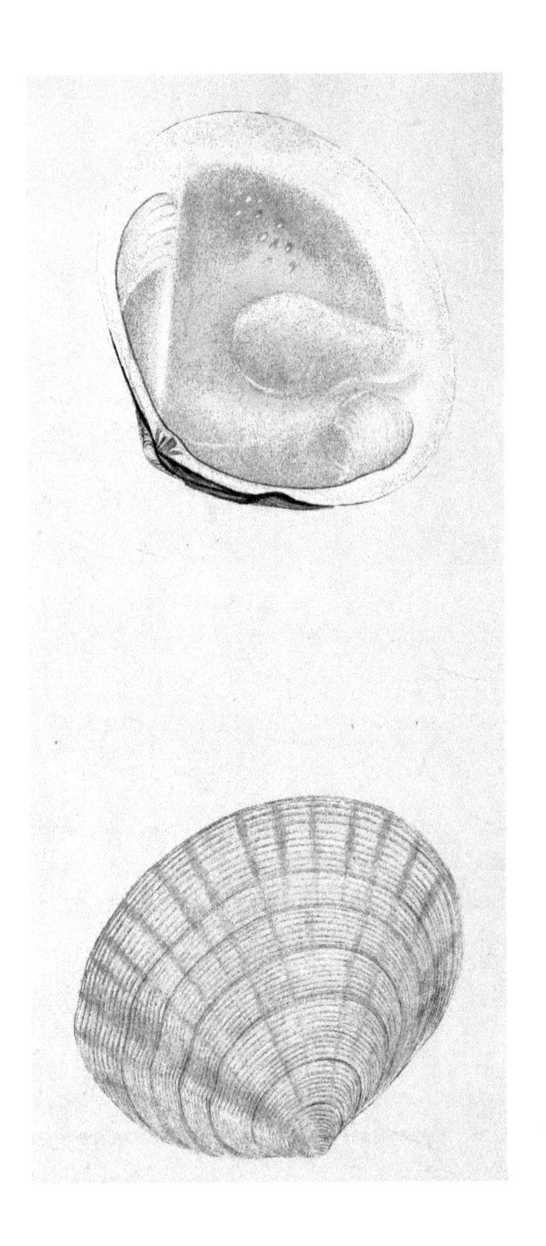

PLATE CIII.

TELLINA RIGIDA.

FLAT AND RIDGED TELLEN.

GENERIC CHARACTER.

The hinge usually furnished with three teeth. Shell generally sloping on one side.

SPECIFIC CHARACTER

AND

SYNONYMS.

Somewhat depressed, subrotund, thick, with numerous transverse thread-like ridges, and a still deeper longitudinal depression near the posterior end.

TELLINA RIGIDA : testa subdepressa subrotunda crassa transversim confertissime striata, postice longitudinaliter magis depressa.

Tellina crassa. *Penn. Br. Zool. p.* 87. *sp.* 28 ?

Pectunculus depressior subrotundus, dense et transversim strigatus. Depressior. *Da Costa Br. Conch. p.* 194. *sp.* 30. *Tab.* 13. *fig.* 4.

Da Costa, who appears to be the only author that describes this shell, says he received it from the coaſt of Cornwall.

PLATE CIII.

This is a thick and heavy shell, though rather transparent; the sides nearly similar, and the beaks almost central. The colour is generally white, with a tinge of yellow on the outside, and some specimens are beautifully radiated with pale pink; the inside is remarkably glossy and finely tinged with yellow, red and orange.

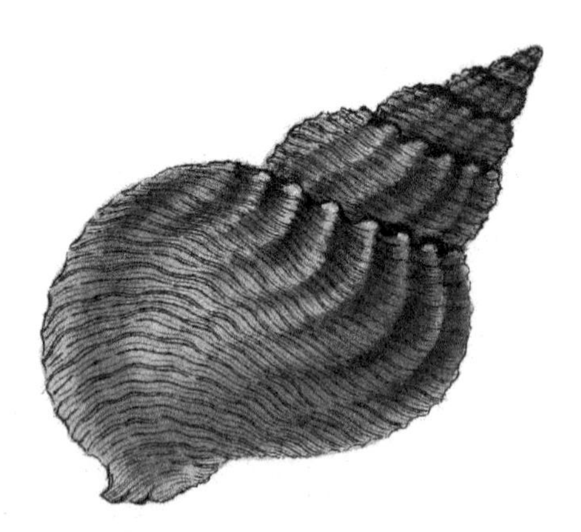

PLATE CIV.

BUCCINUM UNDATUM.

WAVED, OR COMMON WHELKE.

GENERIC CHARACTER.

Aperture oval, ending in a short canal.

SPECIFIC CHARACTER
AND
SYNONYMS.

Shell oblong, coarse, transversely striated, with many curved angles.

BUCCINUM UNDATUM: testa oblonga rudi transversim striata; anfractibus curvato-multangulis. *Gmel. Linn. Syst. Nat. T.* 3. *p.* 3492. *sp.* 93.—*Faun. Suec.* 2263.

Buccinum crassum rufescens, striatum et undatum. *List. H. An. Angl. p.* 156. *tit.* 2. *tab.* 3. *fig.* 2.—Et Bucc. tenue, læve, striatum et undatum. *Id. p.* 157. *tit.* 3. *tab.* 3. *fig.* 3.—Bucc. brevi rostrum tenuiter striatum, pluribus undatis sinubus distinctum. *List. H. Conch. tab.* 962. *fig.* 14.—Et Bucc. brevi rostrum magnum, tenue, leviter striatum. *Id. tab.* 962. *fig.* 15. 15. *a*—*Id. Exerc. Anat. Alt. p.* 68.

PLATE CIV.

Rough, and our most common whelke. *Dale Harw. p.* 382. *No.* 3. 4.

Buccinum undatum, waved, *Penn. Brit. Zool. No.* 90. *pl.* 73.

Buccinum striatum, striated. *Penn. Br. Zool. No.* 91. *pl.* 74.

Buccinum canaliculatum medium vulgare rufescens striatum, pluribus costis undatis distinctum. Vulgare *Da Costa. Br. Conch. p.* 122. *sp.* 73. *tab.* 6. *fig.* 6. 6.

This is the common Whelke of English conchologists, and is sometimes brought to the markets as an article of food. The brown ones are this common sort, for as Linnæus observes, those that are brown fasciated with white or blue are scarce ; the former of those varieties is figured in the annexed plate.

Lister, Pennant and other authors have considered the striated variety of this Shell as a distinct species; it is certainly destitute of those prominent ribs or knobs which is so conspicuous in this Shell in general, but the transitions from the striated kind to those with knobs is so gradual and easy to be traced that we must coincide with Linnæus and Da Costa who admit them barely as varieties.

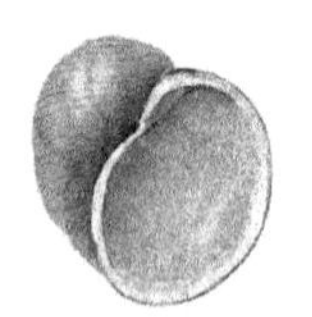

PLATE CV.

HELIX LÆVIGATA.

SMOOTH SNAIL.

GENERIC CHARACTER.

Aperture of the mouth contracted and lunulated.

SPECIFIC CHARACTER AND SYNONYMS.

Imperforate, pellucid, roundish, of two wreaths: the first very large, the second small, obtuse, and placed laterally.

HELIX LÆVIGATA: testa imperforata obovata obtusissima pellucida lævissima. *Gmel. Linn. Syst. Nat. T. I. p.* 6. *p.* 3663. *sp.* 148.

Helix lævigata. Smoothed. *Penn. Br. Zool. T.* 4. *t.* 86. *f.* 139. *Testa M. rar. f.* 17. *Chemn. f.* 1590. 9.

A rare Shell, found on the Kentish coast, and on the beach at Studland, Dorsetshire. Communicated by the Rev. T. Rackett.

Pennant considers this as a fresh water Shell, saying it inhabits ponds. Gmelin is silent respecting its habitation.

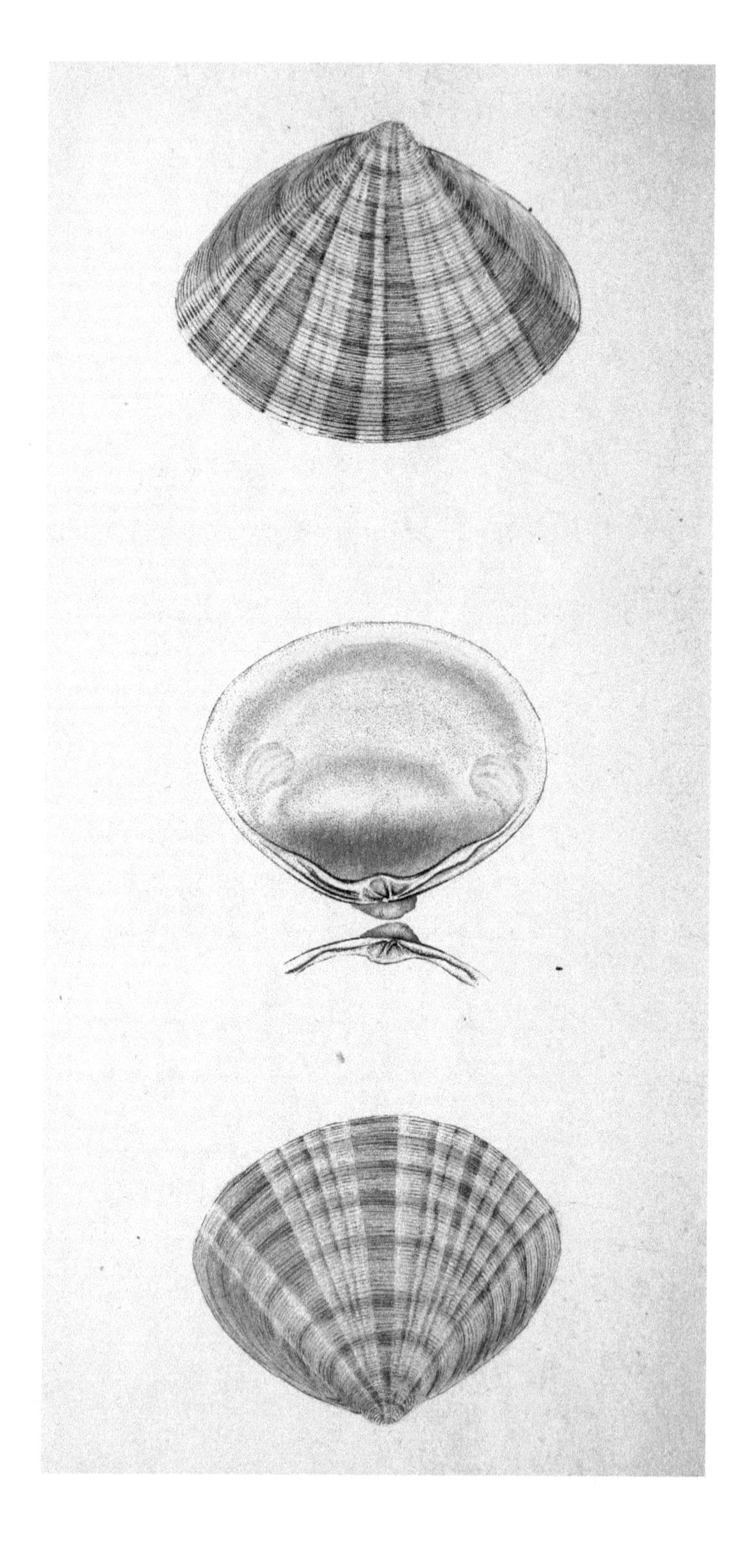

PLATE CVI.

MACTRA STULTORUM.

RAYED MACTRA.

GENERIC CHARACTER.

Animal a Tethys. Bivalve, sides unequal. Middle tooth complicated, with a little groove on each side; lateral teeth remote.

SPECIFIC CHARACTER

AND

SYNONYMS.

Shell semi-transparent, smooth, with faint radiations; within purplish.

MACTRA STULTORUM: testa subdiaphana lævi obsolete radiata, intus purpurascente, vulva gibba. *Gmel. Syst. Nat. T.* 1. *p.* 6. *p.* 3258. *sp.* 11.

Pectunculus triquetrus ex flavo radiatus. *List. H. Conch. tab.* 251. *fig.* 85.

Mactra stultorum, Simpleton. *Penn. Br. Zool. No.* 42. *tab.* 52. *fig.* 42.

Trigonella tenuis admodum concava ferrugineo-cinerea radiata. Radiata. *Da Costa Br. Conch. p.* 196. *sp.* 32.—*Tab.* 12. *fig.* 3. 3.

H 2

PLATE CVI.

This Shell, we observed in plenty, on the sandy shores of South Wales, and particularly on those of Pembrokeshire. It is also found on the western coasts ; at Highlake in Cheshire, near Liverpool ; at the mouth of the river Mersey ; and on the coast of Aberdeenshire and other shores of Scotland.

The general colour of the outside is a kind of milky white, delicately radiated with brown ; within, the young shells are tinged with reddish brown, the old ones with violet.

1

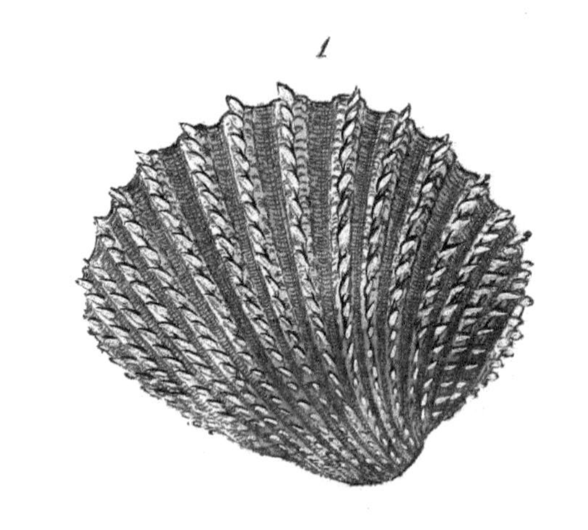

2

PLATE CVII.

FIG. I.

CARDIUM ECHINATUM.

THORNY COCKLE.

GENERIC CHARACTER.

Two teeth near the beak; and another remote one, on each side of the Shell.

SPECIFIC CHARACTER

AND

SYNONYMS.

Shell somewhat heart shaped, ribs prominent with a carinated ridge beset with spines along the middle.

Cardium Echinatum: testa subcordata sulcata: costis carinatis aculeatis. *Linn. Gmel. Syst. Nat. T. I. p.* 6. *p.* 3247. *sp.* 8.

Pectunculus orbicularis fuscus, striis mediis muricatis. *List. H. Conch. tab.* 324. *fig.* 161.

Cardium Echinatum. *Penn. Brit. Zool. No.* 38.

Cardium orbiculare, costis circiter viginti echinatis, spinis hamatis. Echinatum. *Da Costa Brit. Conch. p.* 176. *Tab.* 14. *fig.* 2.

H 3

Dead and worn Shells of this species are found on several of the British coasts in plenty.

It is an elegant shell though the colours are in general obscure: within it is white, without of a pale brown sometimes marked with transverse bands of rust colour.

FIG. II.

CARDIUM TUBERCULATUM.

TUBERCULATED COCKLE.

SPECIFIC CHARACTER

AND

SYNONYMS.

Shell somewhat heart-shaped, ribs obtuse, knotty, transversely striated.

CARDIUM TUBERCULATUM: testa subcordata: sulcis obtusis nodosis transversim striatis. *Linn. Gmel. Syst. T. I. p. 6. p.* 3248. *sp.* 11.

Gmelin mentions several varieties of this species. It has been sometimes considered as the Cardium rusticum.

Found on the coast of Dorsetshire, is scarce, and not before described as a British Shell.

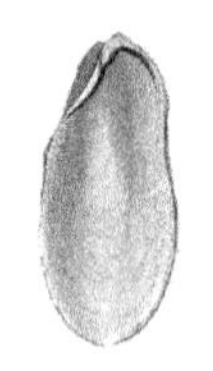

PLATE CVIII.

MYA DUBIA.

DUBIOUS MYA.

GENERIC CHARACTER.

Animal an ascidia. Shell bivalve, gaping at one end. The hinge for the most part furnished with a thick, strong, broad tooth, not inserted in the opposite valve.

SPECIFIC CHARACTER

AND

SYNONYMS.

Shell fragile, brown, bottom widely gaping; rudiment of a tooth in one valve only.

MYA DUBIA: testa fragili fusca subtus valde hiante valva una edentula altera rudimento dentis.

Mya dubia. *Penn. Br. Zool. p.* 82. 19.

Pennant, who seems to be the only author that describes this shell, says it has the rudiment of a tooth within one shell; with an oval and large hiatus opposite the hinge. Shells brown and brittle, size of a Pistachia nut. Length of a Horsebean, and found near Weymouth.

This Shell is rare, Pennant notes his from the Portland cabinet.

INDEX

TO

VOL. III.

LINNÆAN ARRANGEMENT.

BIVALVIA: CONCHÆ.

I N D E X.

INDEX TO VOL. III.

ACCORDING TO

HISTORIA NATURALIS TESTACEORUM BRITANNIÆ OF DA COSTA.

PART I.

GENUS 2.

* MARINÆ. SEA.

PART II.

UNIVALVIA INVOLUTA.

GENUS 5. BULLA. DIPPER.

INDEX.

PART III.

UNIVALVIA TURBINATA.

GENUS 7. TROCHUS TOP SHELL.

* MARINÆ. SEA.

GENUS 9. HELIX.

** FLUVIATILES. RIVER.

GENUS 10. COCHLEA SNAILS.

* TERRESTRES. LAND.

** FLUVIATILES. RIVER.

MARINÆ. SEA.

I N D E X.

GENUS 11. TURBO.

* TERRESTRES. LAND.

** FLUVIATILES. RIVER.

GENUS 13. BUCCINA CANALICULATA. GUTTERED WHELKS.

* MARINÆ. SEA.

GENUS 14. BUCCINA RECURVIROSTRA. WRY-MOUTHED WHELKS.

ORDER 2.

BIVALVES.

GENUS 6. CARDIUM. HEART COCKLE.

* FLUVIATILES. RIVER.

I N D E X.

ALPHABETICAL INDEX TO VOL. III.

INDEX.

END OF VOL. III.

Printed by Bye and Law, St. John's-Square, Clerkenwell.

THE

NATURAL HISTORY

OF

BRITISH SHELLS,

INCLUDING

FIGURES AND DESCRIPTIONS

OF ALL THE

SPECIES HITHERTO DISCOVERED IN GREAT BRITAIN,

SYSTEMATICALLY ARRANGED

IN THE LINNEAN MANNER,

WITH

SCIENTIFIC AND GENERAL OBSERVATIONS ON EACH.

VOL. IV.

By E. DONOVAN, F.L.S.

AUTHOR OF THE NATURAL HISTORIES OF
BRITISH BIRDS, INSECTS, &c. &c.

LONDON:

PRINTED FOR THE AUTHOR,

AND FOR

F. AND C. RIVINGTON, No 62, ST. PAUL'S CHURCH-YARD;

BY BYE AND LAW, ST. JOHN'S SQUARE, CLERKENWELL.

1802.

THE

NATURAL HISTORY

OF

BRITISH SHELLS.

PLATE CIX.

MUREX CARINATUS.

CARINATED MUREX.

GENERIC CHARACTER.

Spiral, rough. The aperture ending in a strait, and somewhat produced gutter or canaliculation.

SPECIFIC CHARACTER

AND

SYNONYMS.

Tail patulous: Shell oblong, of six spires, with two smooth spiral ridges; first spire ventricose. Aperture semi-circular.

Murex Carinatus: testa patulo-subcaudata oblonga: anfractibus sex lævibus bicarinatis; primo ventricoso, apertura semicirculari.

A 2

PLATE CIX.

Murex carinatus, angulated. With five or six spires, the body ventricose: the spires rising into angulated ridges. The aperture semicircular. Length near four inches. From the Portland Cabinet. *Penn. Br. Zool. T. 4. p. 123. sp. 96.*

The shell figured in the annexed Plate is unique; it formerly belonged to the late Duchess of Portland, by whose permission Mr. Pennant described it in the British Zoology. This author has given two figures of it, one in Plate 77, and the other in the Frontispiece of the fourth volume.

The existence of this species being only proved by a solitary specimen, various conjectures have arisen amongst Conchologists respecting it. Some have been inclined to admit it as an undoubted species, and others as a mere accidental variety of growth of the Linnæan Murex Antiquus. How far we may be authorized to abide by the former opinion must rest with the critical Naturalist.

To argue that it cannot be a distinct species, because only one shell of the kind has been hitherto found, is absurd; since the existence of many other species has been asserted upon the evidence of a single specimen only, and its relation to Murex antiquus is not so obvious as might be at first imagined. It certainly approaches it in the general outline, but the ridges of Murex Antiquus is most completely raised into tubercules, whereas those of Carinatus are perfectly smooth and even, nor is there that strict correspondence in the angulations of the contour in general that should induce us to consider it a variety of Murex Carinatus.

PLATE CIX.

In deciding a question of some moment to the English Conchologist, it has been thought advisable to give an additional Plate of Murex Antiquus, by which the difference between the two shells may be more easily discriminated. We must however observe, that the latter is not absolutely known as a British shell; it is a native of the North Seas, and has been supposed to inhabit some of the remote northern islands of the British dominions. The Murex Antiquus of Pennant is a very different shell, and by no means allied to that of Linnæus, whose name it bears.

It is now uncertain from what part of our coast the Duchess of Portland received this shell; Pennant is silent in this respect, but we cannot dispute that her Grace received it as a British shell, since it was inserted upon her authority in the British Zoology.

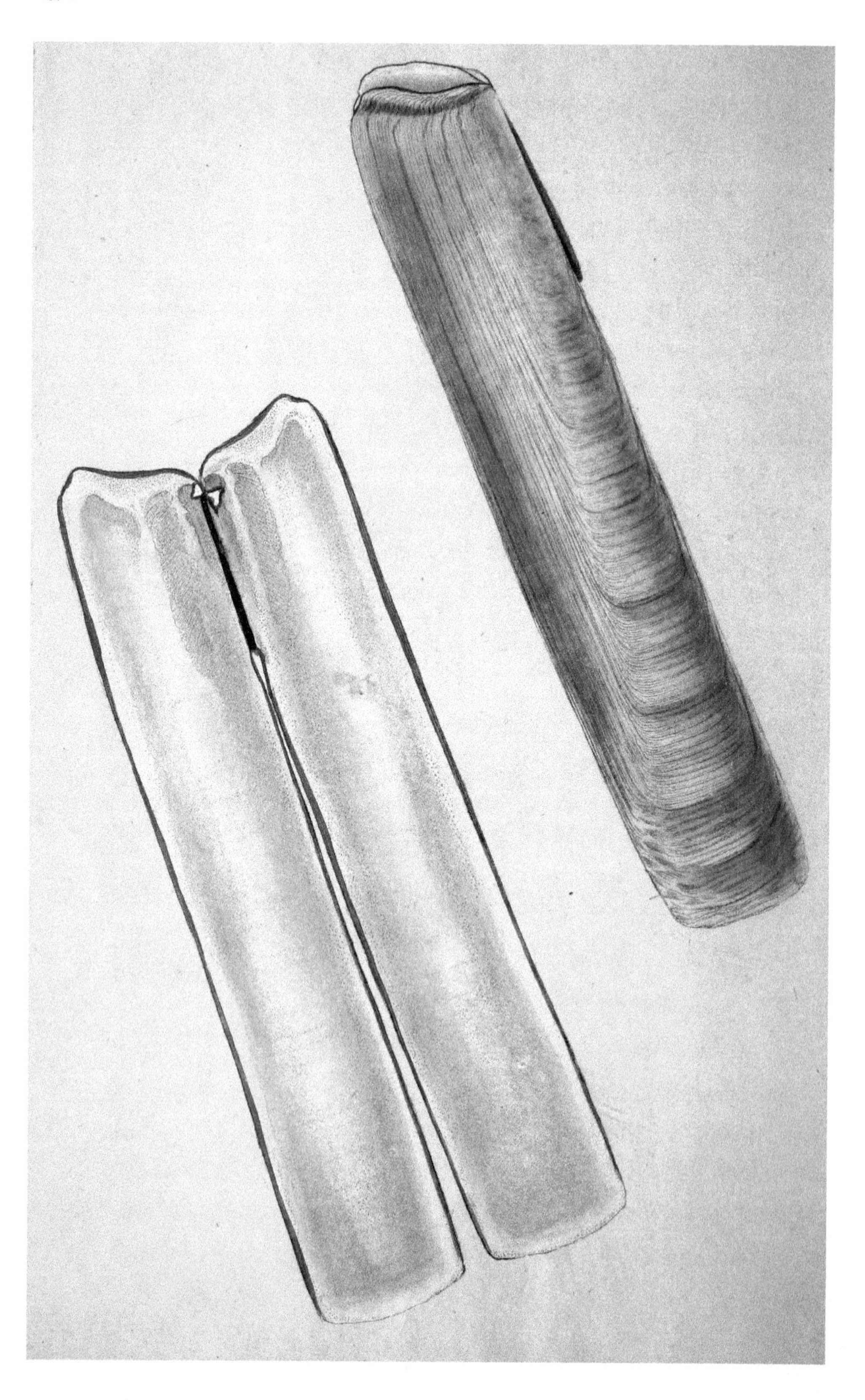

PLATE CX.

SOLEN MARGINATUS.

MARGINATED RAZOR SHELL.

GENERIC CHARACTER.

Bivalve, with equal valves, oblong, open at both ends. At the hinge a subulated tooth turned back, often double; not inserted in the opposite shell. Animal an ascidia.

SPECIFIC CHARACTER
AND
SYNONYMS.

Shell straight, of equal depth, a single tooth in each valve.

SOLEN MARGINATUS: testa lineari recta marginata, valvulis un-dentatis.

Solen Vagina, Sheath, *Penn. Br. Zool. p.* 83. *No.* 21.

Some Conchologists imagine that Da Costa has confounded this species with his Solen Siliqua, but it is more probable that he had never met with it, or the character of the teeth at the hinge could not have escaped his notice. In its general appearance it is not unlike Solen Siliqua, but has one end marginated, and only a single tooth in each valve; on the contrary Solen Siliqua has two teeth in one valve, and one in the other; the single one being inserted between the two others when the shell is shut.

This is very scarce. Pennant says it inhabits Red Wharf, Anglesea, —This is not Solen Vagina of Linnæus, as Pennant describes it.

PLATE CXI.

TROCHUS TERRESTRIS.

LAND TOP SHELL.

GENERIC CHARACTER.

Animal a slug. Shell conic, aperture nearly triangular.

SPECIFIC CHARACTER.

Rather conic, whitish, with a spiral brown streak along the middle of the wreaths.

TROCHUS TERRESTRIS: testa subconica albida anfractibus linea media fusca.

Trochus Terrestris, Land. *Penn. Brit. Zool. No.* 108. *tab.* 80. *fig.* 108.

Trochus Terrestris tertius. *Da Costa Br. Conch. p.* 36. *C.*

Pennant describes this new British species of Land Trochus upon the authority of Mr. Hudson, who discovered it upon the Mountains of Cumberland. Da Costa therefore places it as a distinct species, but expresses some doubt whether it may not be the same Land Trochus as Dr. Lister found in the moss at the roots of the large trees in Burwell woods, in Lincolnshire, and to which the shell found by Mr. Morton, in Morsley wood, Northamptonshire, bears great affinity. Dr. Lister's shell had six or seven wreaths, and Mr. Morton's only five. *Mort. Northampt. ch.* 7. *p.* 415.

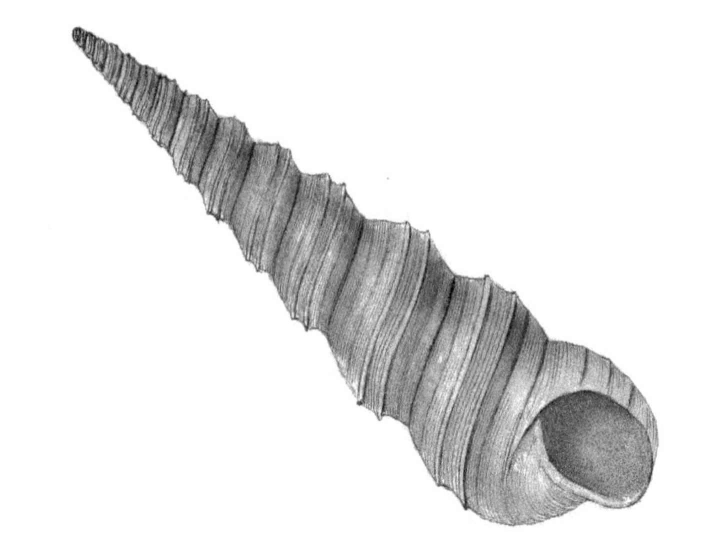

PLATE CXII.

TURBO DUPLICATUS.

TWO RIDGED SCREW SHELL.

GENERIC CHARACTER.

Animal Limax. Univalve, spiral, or of a taper form. Aperture somewhat compressed, orbicular, entire.

SPECIFIC CHARACTER

AND

SYNONYMS.

Shell slender, with two sharp prominent spiral ridges.

Turbo Duplicatus: spiræ anfractibus carinis duabus acutis. *Gmel. Linn. Syst. p.* 3607. *sp.* 79.

Buccinum crassum, duobus acutis, & inæqualiter altis striis in singulis duodecim minimum spiris donatum. An. Buccinum striatum σαλπιγξ Fab. Columnæ? *List H. An. Angl. p.* 160. *tit.* 7. *tab.* 3. *fig.* 7.

Turbo duplicatus, doubled. *Penn. Br. Zool. No.* 112. *tab.* 81. *fig.* 112.

Strombiformis major rubro lutescens aut pullus: anfractibus duabus carinis sive striis acutis insignitis. Bicarinatus. s. Torcular. *Da Costa. Br. Conch. p.* 110. 44.— *Tab.* 6. *fig.* 3.

PLATE CXII.

This species is admitted as a British shell upon the authority of Dr. Lister, who says he had purchased them of the Scarborough fishermen. Dr. Lister had not seen any of them alive, and concludes it must be a pelagian shell, or one of those which live far from the shores.

As Pennant had inserted this species in his *British Zoology*, upon this authority, Da Costa was unwilling to omit it in his *British Conchology*, yet he observes, that it is not improbable Dr. Lister was imposed upon by the fishermen, for the shell is generally believed to be a native of the East Indies ; some consider it as a West-Indian, and others as an European species.

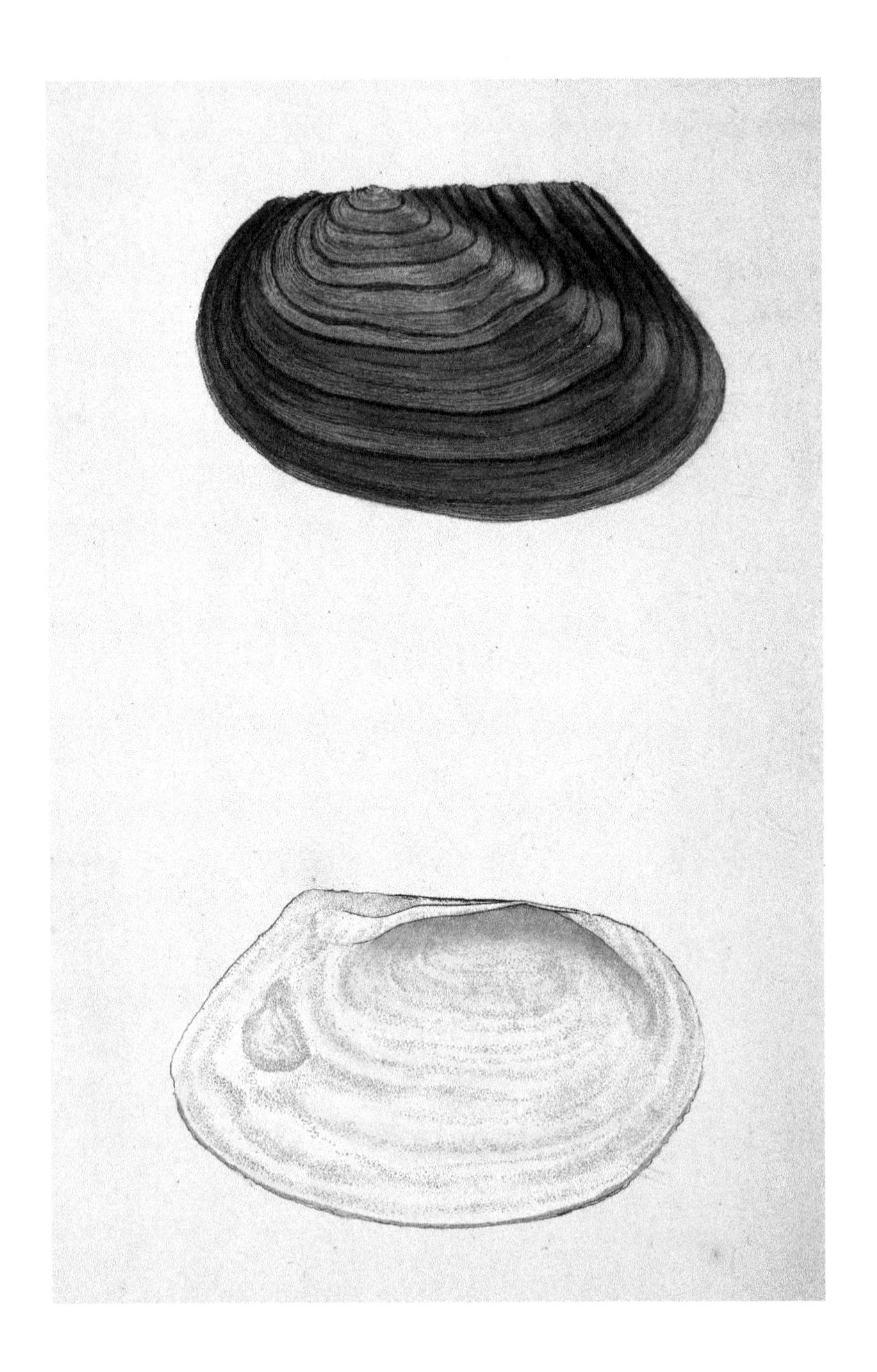

PLATE CXIII.

MYTILUS ANATINUS.

SMALL HORSE MUSCLE.

GENERIC CHARACTER.

The hinge toothless, and consists of a longitudinal furrow.

SPECIFIC CHARACTER

AND

SYNONYMS.

Shell oval, rather compressed, brittle, margin membranaceous, beaks decorticated.

MYTILUS ANATINUS: testa ovali compressiuscula fragilissima margine membranceo, natibus decorticatis. *Gmel. Linn. Syst. Nat. p.* 3355. *sp.* 16.

Musculus latus, testa admodum tenui, ex fusco viridescens, interdum rufescens, &c. *List. H. An. Angl. p.* 146. *tit.* 29. *tab.* 2. *fig.* 29.

Musculus tenuis minor latiusculus. *App. H. An. Angl. p.* 10. *tit.* 30. *tab.* 1. *fig.* 2.—*App. H. An. Angl. in Goed. p.* 13. *tit.* 30. *tab.* 1. *fig.* 2.

Mytuli majores à nostratibus. Horse muscles. *Merret. Pin. p.* 193.

Mytilus anatinus Duck. *Penn. Br. Zool. No.* 79. *tab.* 68. *fig.* 79.

Mytilus fluviatilis minor. Anatinus, *Da Costa Br. Conch. p.* 215. *sp.* 47. *tab.* 15. *fig.* 2.

PLATE CXIII.

This species bears much resemblance to Mytilus Cygneus, but differs in being only about half the size, is more compressed and oblong, of a clearer green colour, and the cartilage side extending in a straight line to an acute angle, like a fin, and thence continuing in an oblique line towards the bottom, where it is rounded.—Extremely common in rivers and stagnant waters.

Pennant's shell is much broader in proportion than our specimens.

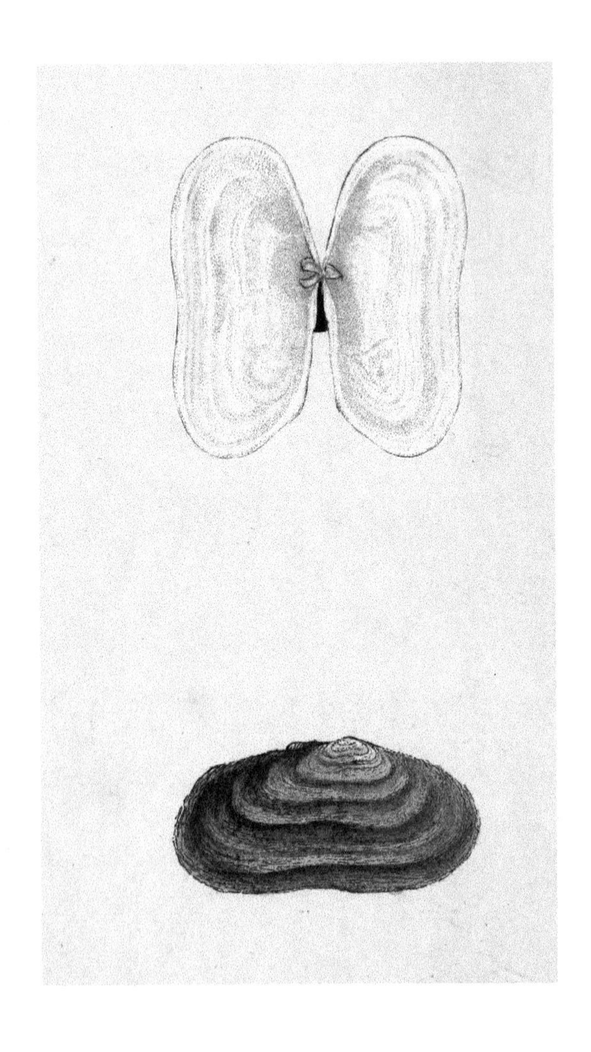

PLATE CXIV.

SOLEN ANTIQUATUS.

ANTIQUATED SOLEN, OR RAZOR SHELL.

GENERIC CHARACTER.

Bivalve, with equal valves, oblong; open at both ends. At the hinge a subulated tooth turned back, often double; not inserted in the opposite shell. Animal an Ascidia.

SPECIFIC CHARACTER AND SYNONYMS.

Oval oblong, semipellucid, lower margin sinuous in the middle.

SOLEN CHAMA-SOLEN: testa ovali-oblonga subpellucida, sinuosa. *Da Costa. Br. Conch. p.* 238. *sp.* 62.

Chama angustior, ex altera parte sinuosa. *List. H. Conch. tab.* 421. *fig.* 265.

Solen Cultellus, Kidney. *Penn. Brit. Zool. No.* 25. *tab.* 46. *fig.* 25.

This is perhaps the rarest species of the Solen genus found upon the British coasts, and as Pennant observes, seems to connect the Solen with the Mya genera. It borders on the Chama of Da Costa, who therefore calls it Solen Chama-Solen. Pennant notes it from Weymouth, and Da Costa received it from the shores of Dorsetshire and Hampshire.

Pennant has mistaken this for a very distinct shell, described by Linnæus, under the name of Solen Cultellus.

PLATE CXV.

VENUS CANCELLATA.

MEMBRANACEOUS VENUS.

GENERIC CHARACTER.

Bivalve. Hinge furnished with three teeth; two near each other, the third divergent from the beaks.

SPECIFIC CHARACTER

AND

SYNONYMS.

Somewhat heart shaped with remote transverse membranaceous ridges; a cordiform depression on the slope under the beaks.

VENUS CANCELLATA: testæ striis transversis membranaceis remotis, Ano cordato. *Gmel. Linn. Syst. p.* 3270. *sp.* 8.

Pectunculus strigis transversis remotis, acutis, membranaceis, donatus Membranaceous. *Da Costa Br. Conch. p.* 193. *sp.* 29. *tab.* 13. *fig.* 4. right hand.

Da Costa described this shell from a specimen in the collection of the late Dr. Fothergill. It is from the Western coast.

The shell figured by Pennant, No. 48. A. Pl. 48, as a Worn shell of Venus Erycina, is probably of this species; for it seems entirely destitute of the longitudinal undulations that decussate the transverse ridges in Venus Erycina.

116

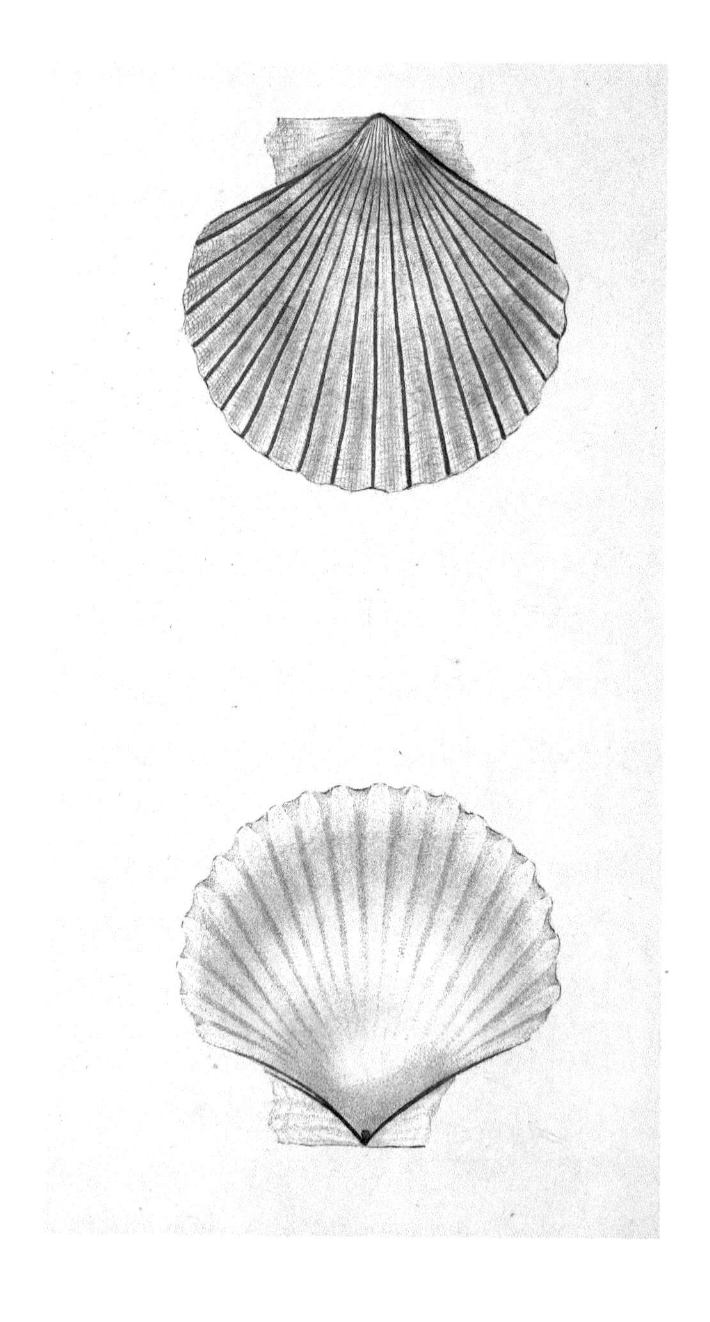

PLATE CXVI.

OSTREA LINEATA.

LINEATED SCALLOP.

GENERIC CHARACTER.

Animal a Tethys. Shell bivalve, unequal. The hinge without a tooth, having a small oval cavity.

SPECIFIC CHARACTER

AND

SYNONYMS.

Valves nearly equal, thin: one white, the other marked with a single purple line down each rib.

OSTREA LINEATA: testa subæquivalvi tenui, valva una alba, alterius singulis costis linea purpurascenti.

Pectunculus, mediocris, fere æquivalvis, tenuis, valva una alba, altera vero cum linea purpurascente in summitate unaquæque costæ. Lineatus. *Da Costa Br. Conch. p.* 147. *sp.* 4. *Tab.* 10. *fig.* 8.

This elegant shell seems to be described only by Da Costa; he says he was informed it had been fished up about Weymouth, in Dorsetshire, but had only seen it from Cornwall. We have it from the coast of Devonshire also, from which it appears an inhabitant of the western coast in general, though it is very rarely met with.

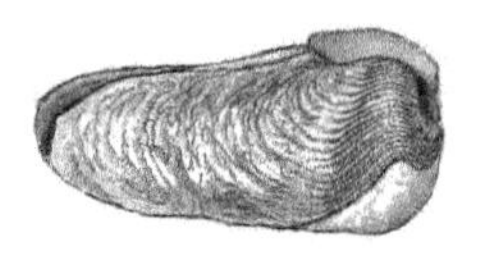

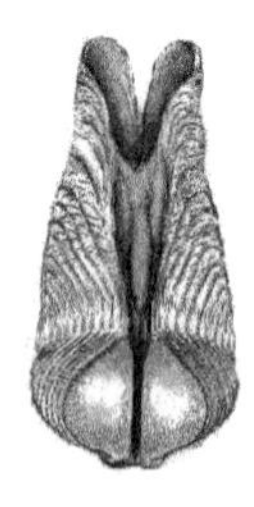

PLATE CXVII.

PHOLAS STRIATA.

STRIATED PIDDOCK.

GENERIC CHARACTER.

Animal ascidia. Shell bivalve, opening wide at each end, with several lesser valves at the hinge. The hinges folded back and connected by a cartilage.

SPECIFIC CHARACTER

AND

SYNONYMS.

Shell ovate, with numerous striæ.

PHOLAS STRIATA: testa ovata multifarium striata. *Gmel. Linn. Syst. p.* 3215. *sp.* 3.
Act. Angl. 55. *t.* 1. *f.* 1, 2, 3, 4.
Gualt. Test. t. 105. *F.*

Extremely rare as a British species, and not noticed by either Pennant or Da Costa. In the collection of the Rev. T. Rackett.

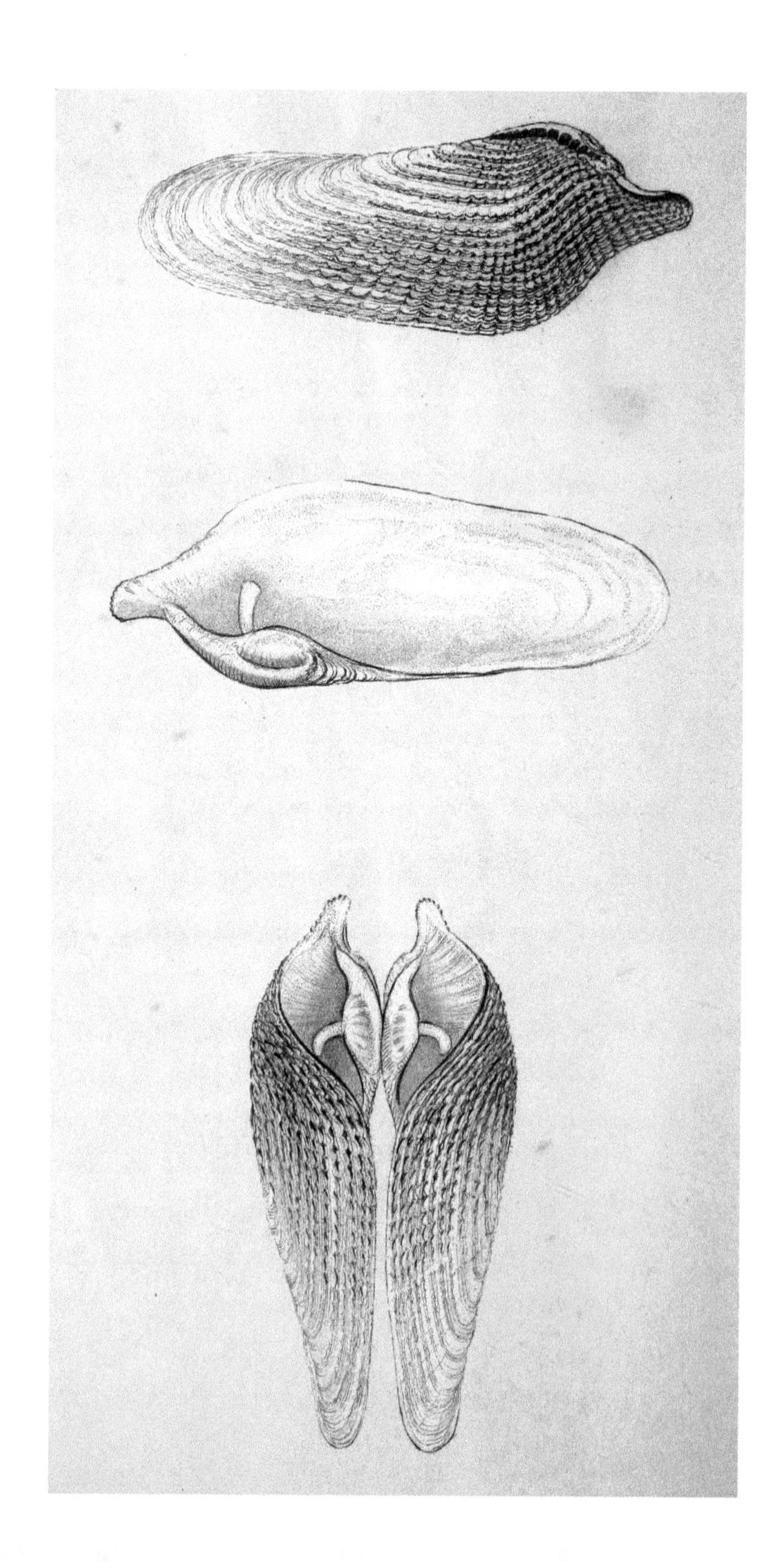

PLATE CXVIII.

PHOLAS DACTYLUS.

PRICKLY PADDOCK.

GENERIC CHARACTER.

Animal ascidia. Shell bivalve, opening wide at each end, with several lesser valves at the hinge. The hinges folded back and connected by a cartilage.

SPECIFIC CHARACTER

AND

SYNONYMS.

Shell oblong, striated transversely, and reticulated on the upper part with little spines.

PHOLAS DACTYLUS: testa oblonga hinc reticulato-striata. *Gmel. Linn. Syst. p.* 3214.

Pholas rostratus major diepensis vulgò Gallice Piteau dictus. *App. H. An. Angl. in Goed. p.* 37. *tab.* 2. *fig.* 3.—Pholas striatus, sinuatus ex altera parte. *Hist. Conch. tab.* 433. *fig.* 276.—Pholas alte striatus, ex altera parte sinuatus, eadem mucronatus, Hist. nost. Conch. Anglice Piddocks, Gallicæ Pitau; earumque piscatories pitau quieres. Exercit. *Anat.* 3. *p.* 88. *tab.* 7. *fig.* 1, 2. Pholas angustius; oblong Pierce stone or Pholade. *Petiv.*

PLATE CXVIII.

Gaz. tab. 79. *fig.* 10.—Piddocks. ***Dale Harw.*** p. 389. Pholas Dactylus, Dactyle. ***Penn. Br. Zool.*** *p.* 76. *sp.* 10.

Pholas angustius striatus & veluti aculeatus. Muricatus. ***Da Costa*** *Br. Conch. p.* 244. *sp.* 65. *tab.* 16. *fig.* 2. 2.

This species burrows or pierces into rocks, where it forms large cylindrical cavities. It is not uncommon on many of our coasts, and is sometimes eaten; it is in season in the Spring.

119

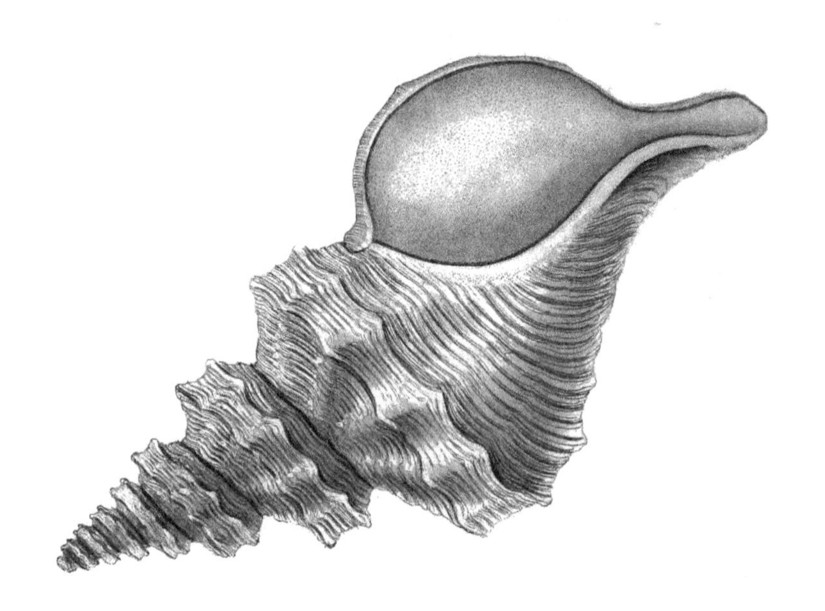

PLATE CXIX.

MUREX ANTIQUUS.

ANTIQUATED MUREX.

GENERIC CHARACTER.

Spiral, rough. The aperture ending in a strait, and somewhat produced gutter or canaliculation.

SPECIFIC CHARACTER.

Tail patulous: Shell oblong, of eight spires: spiral ridges tuberculated.

MUREX ANTIQUUS: testa patulo-caudata oblonga: anfractibus octo teretibus. *Faun. Suec.* 2165.—*Gmel. T.* 1, *fig.* 6. *p.* 3546.

In the description of Plate CIX. our reafon for considering this and Murex Carinatus as two distinct species, are briefly stated; and the difference, it is presumed, will be farther apparent on comparing the two shells figured in that, and the annexed Plate.

Reversed shells of this species have been sometimes found. It is an inhabitant of the northern parts of Europe.

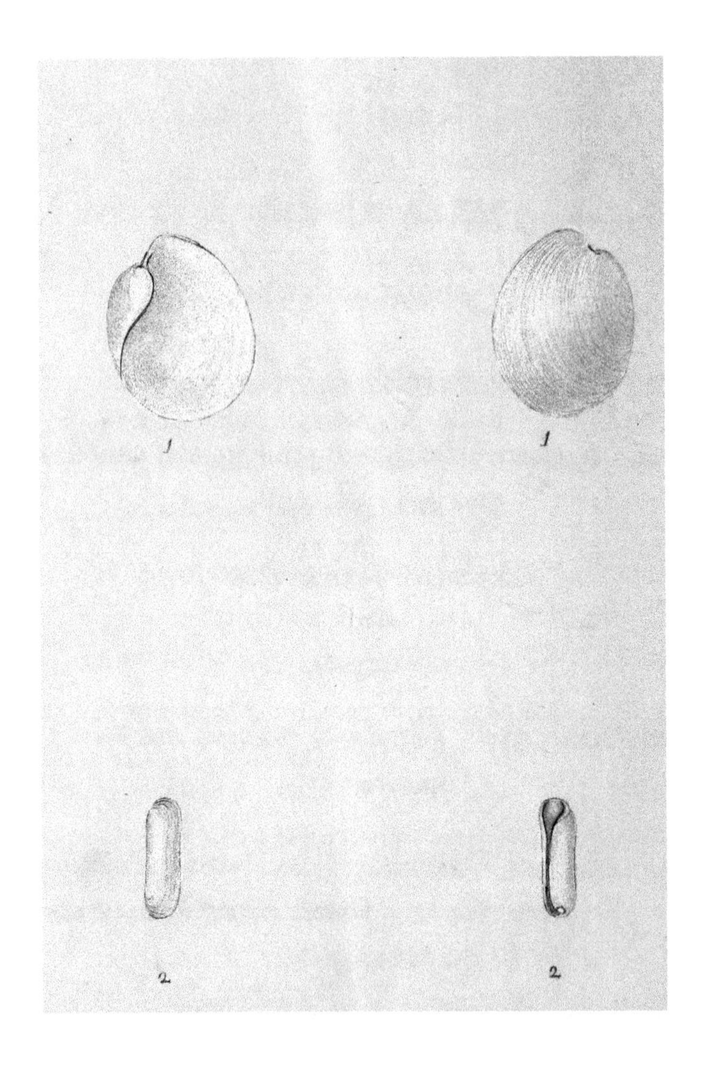
1
1
2
2

PLATE CXX.

FIG. I.

BULLA APERTA.

OPEN BULLA OR BUBBLE.

GENERIC CHARACTER.

Shell suboval, Aperture oblong, very patulous, and smooth or even. One end convoluted.

SPECIFIC CHARACTER
AND
SYNONYMS.

Shell almoſt entirely open, subrotund, pellucid, and faintly striated transversely.

BULLA APERTA: testa subrotunda pellucida transversim substriata tota hiante. *Gmel. Linn. Syst. Nat. T.* 1. *p.* 6. *sp.* 3424. *sp.* 8.

Bulla pellucida, fragilissima, tota hians, s. apertura amplissima. *Da Costa. Brit. Conch. p.* 30. *sp.* 16.

Da Costa says, all the shells of this species he knew, were fished up near Weymouth in Dorsetshire, and not any where else on the British coast; they are even not frequent there; so that it seems a rare as well as curious shell.—To this we may add, that they are

D 2

rare only because they are local: in one part of the sandy bay of Caermarthen, below Tenby, we found them in abundance.

It is called "the Bubble" by this writer; who observes that it exactly resembles a bubble or bladder of water. The aperture is so extremely large that the whole shell lies open to view. The contour is somewhat oval, and slightly involuted; and the shell is not umbilicated.

This is certainly not Bulla patula of Pennant (Brit. Zool. No. 85. A.) as Da Costa and Gmelin imagine. The figures in that work are sometimes calculated to mislead the most attentive; but as we are in possession of the shell Pennant describes, we can venture to say the two former writers are mistaken. The species Aperta was unknown to Pennant, and the shell he figured from the Portland Cabinet, under the specific name of Patula, is extremely rare.

FIG. II.

BULLA CYLINDRICA.

NARROW BULLA.

SPECIFIC CHARACTER.

Shell cylindric, smooth, white, and thin.

BULLA CYLINDRICA: testa cylindrica lævi alba tenuissima. *Gmel.* *T.* 1. *p.* 6. *p.* 3433. *sp.* 38.

PLATE CXX.

This is a very scarce species on our coasts, and approaches so nearly to Bulla pallida of Da Costa (Voluta pallida. *Linn.*) described and figured in Plate LXVI. of this Work, that it may easily be confounded with it, unless the two shells be compared. Bulla Cylindrica differs in being rather more compressed, and has the pillar-lip perfectly smooth; while on the contrary the other has plaits or wrinkles upon this part as before observed: a circumstance that constitutes one character of the Voluta genus in the system of Linnæus.

In the description of the Voluta pallida we were led to think with Da Costa, that the shell figured by Pennant in the British Zoology, No. 85. A. might be of the same species. Since that time we have been favoured with the specimen figured in the annexed Plate; and as it seems to correspond more clearly with Pennant's shell than the other, there can be no impropriety in removing the reference from the former to the prefent species. Gmelin takes not the slightest notice of this figure of Pennant; so that we must remain ignorant of his opinion respecting it. And it is certain the shell before us was altogether unknown to Da Costa.

D 3

121

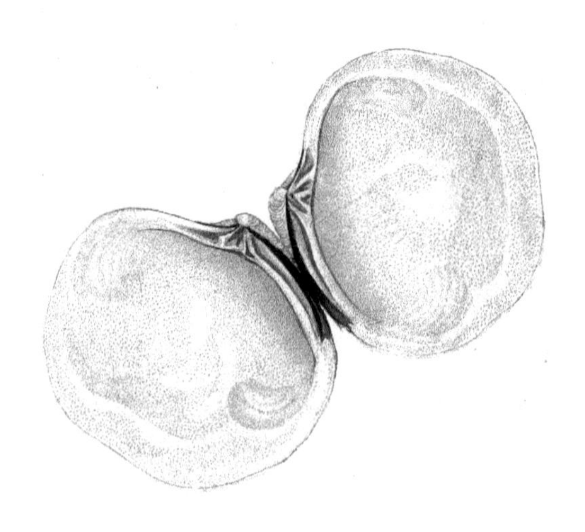

PLATE CXXI.

VENUS UNDATA.

WAVED VENUS.

GENERIC CHARACTER.

Bivalve. Hinge furnished with three teeth; two near each other, the third divergent from the beaks.

SPECIFIC CHARACTER

AND

SYNONYMS.

Shell orbiculated, convex, thin, transversely marked with very fine striæ, and waved at the margin.

VENUS UNDATA: testa orbiculata convexa tenui transversim subtilissime striata margine undata.

VENUS UNDATA, *waved.*—With thin, convex, orbiculated shells, of a white colour, tinged with yellow, and marked with thin concentric *striæ*; waved at the edges. *Penn. Brit. Zool.* 4. *sp.* 51.

There can be no doubt that the shell figured and defcribed by Pennant in the fourth volume of his British Zoology, *No.* 51, is of

D 4

the ſame ſpecies as our shell. That author says it is the size of a hazel nut, from which it appears the ſpecimen he ſaw was a young Shell; the largest of our specimens being of the size represented in the annexed Plate.

This is *Venus lactea* of some cabinets, a name sufficiently expressive of its colour, but having been before called *Undata* by Pennant, we thought it best to retain the name he had given it.

122

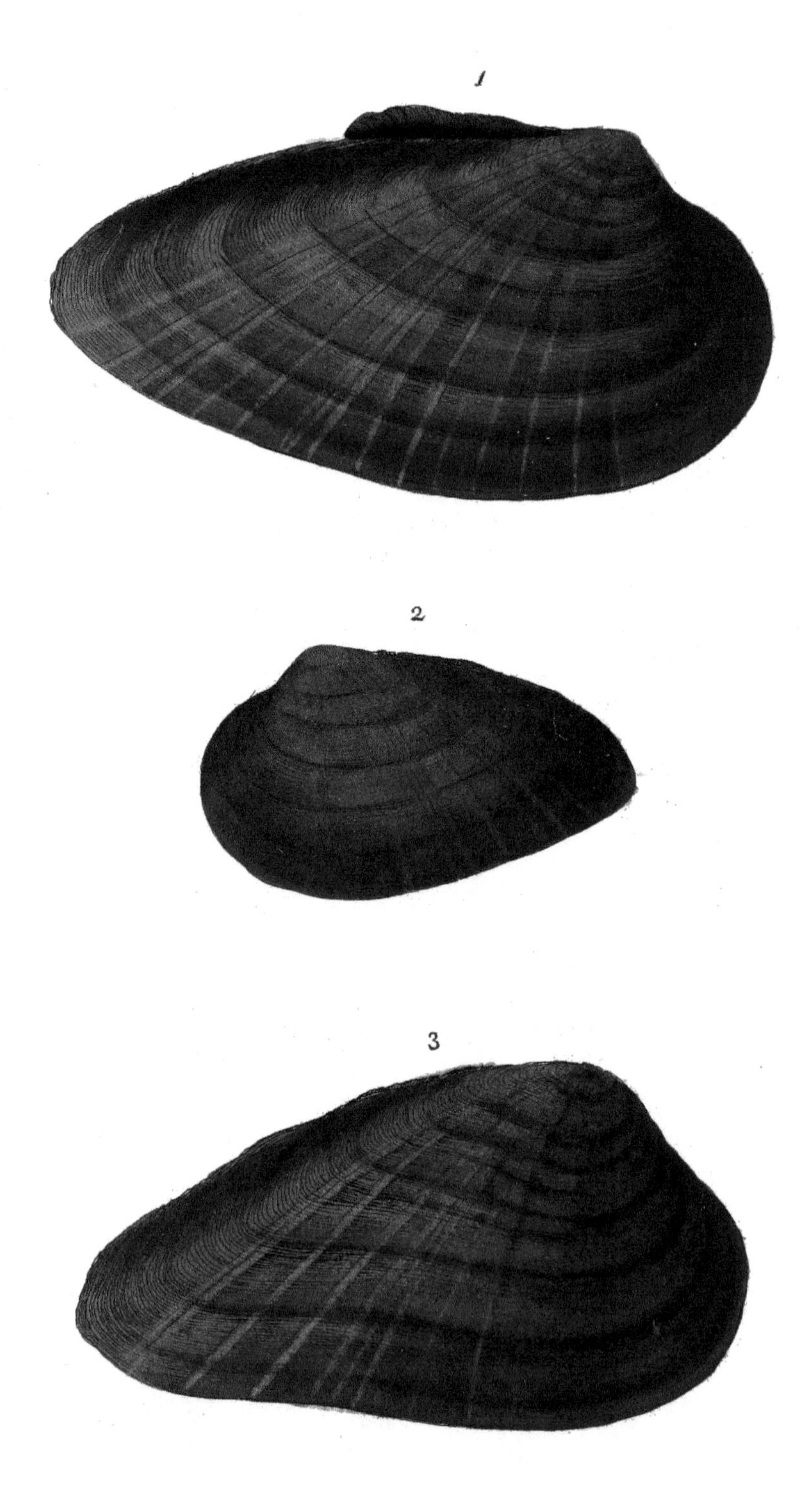

PLATE CXXII.

MYA OVATA.

OVATE MYA.

GENERIC CHARACTER.

Animal an afcidia. Shell bivalve, gaping at one end. The hinge for the most part furnished with a thick, strong, and broad tooth, not inferted in the opposite valve.

SPECIFIC CHARACTER.

Shell oblong-ovate; posterior part roundish, and very slightly gaping; first tooth at the hinge crenulated.

MYA OVATA: testa oblongo-ovata, postice rotundata vix hiante, cardinis dente primario crenulato.

This is the Shell alluded to in the description of Plate 101, under the name of Mya ovata. The difference between it and Mya depressa, as before observed, seems to consist in the present being more ovate, and not depressed across the middle: nor is the gaping at the broadest end so considerable as in the other.

These, we have already remarked, inhabit the same waters as Mya depessa, viz. in the New River, near London, and the Froome in

Somerfetshire. It is a thick, strong and heavy Shell, of a greenish colour, and radiated.—The smallest figure in the annexed plate is a brown coloured specimen of Mya depressa.

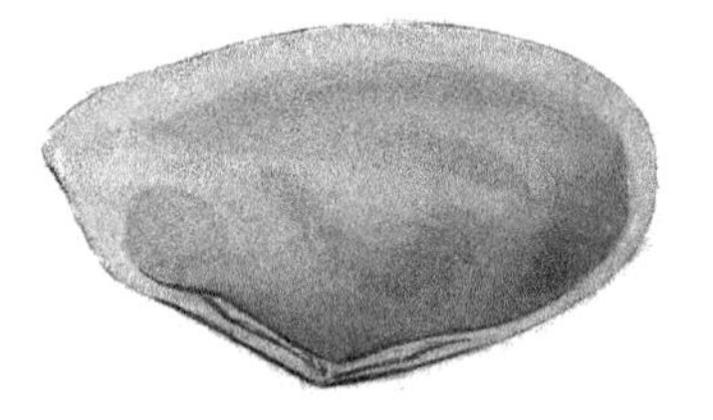

PLATE CXXIII.

TELLINA INÆQUISTRIATA.

UNEQUALLY-STRIATED TELLEN.

GENERIC CHARACTER.

The hinge usually furnished with three teeth. Shell generally sloping on one side.

SPECIFIC CHARACTER.

Shell ovate, compressed and rather flattish, rosy, very finely striated transversely: the striæ fewer and larger at the anterior end.

TELLINA INÆQUISTRIATA: testa ovata compresso-planiuscula rosea subtilissime transversim striata: striis anterius paucioribus majoribusque.

A very rare species of Tellina communicated to Da Costa after his Conchology was published, and therefore not noticed in that work. It has been found by the late Dr. Pulteney we believe on the coast of Dorsetshire.

2

1

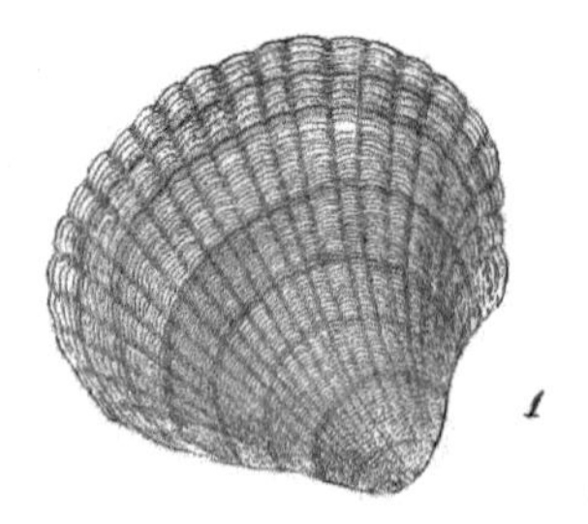
1

PLATE CXXIV.

FIG. I.

ARDIUM EDULE.

COMMON COCKLE.

GENERIC CHARACTER.

Two teeth near the beak; and another remote one, on each side of the shell.

SPECIFIC CHARACTER
AND
SYNONYMS.

Shell antiquated: about twenty-six grooves, with obsolete recurved scales.

CARDIUM EDULE: testa antiquata; sulcis viginti sex obsolete recurvato imbricatis. *Linn.—Gmel. T.* 1. *p.* 6. *p.* 3252. *sp.* 20.

PECTUNCULUS VULGARIS, albidus, subrotundus, circiter viginti-sex striis majusculis at planioribus donatus. *Da Costa, Brit. Conch. p.* 180. *sp.* 19.

Cardium Edule, Edible Cockle, *Penn. Brit. Zool. No.* 41. *tab.* 50. *fig.* 41.

PLATE CXXIV.

The Common Cockle is abundant on all sandy shores: they lurk in the sand, and their hiding-place is known by a little round depressed spot upon the surface. Cockles are in season from autumn till spring: they are a wholesome and palatable food; and thofe from Selfea, near Chichester, are esteemed the most delicious in England.

These Shells vary a little both in shape and colour: the two fpecimens figured on the annexed Plate differ, one being more orbicular than the other. They are generally whitish, sometimes they have a blueish, and sometimes a yellowish tint.

FIG. II.

CARDIUM RUSTICUM?

SPECIFIC CHARACTER.

Shell antiquated, with about twenty remote grooves: the interftices rugged.

CARDIUM RUSTICUM: testa antiquata: sulcis viginti remotis; interstitiis rugosis. *Gmelin*, *T*. 1. *p*. 6. *p*. 3252. *sp*. 23?

Notwithstanding the endless variations to which the shells of the common Cockle are liable, this appears too remote to be admitted as one of them. It passes for Cardium rusticum with some conchologifts, and though it may not strictly agree with that specific description of Gmelin, it approaches nearer to it than to edule. That author notices the affinity *rusticum* bears to the other species; but observes that the grooves are deeper and the ribs fewer, and more convex in rusticum: he remarks also that the latter has a ridge on the anterior margin when the valves are closed, and a narrow depressure behind the beaks, " ano evidente, sed angusto," which is not in the other. The grooves in our Shell are not so deep as " sulcis profundioribus" implies, but they are both deeper and wider than in the common sort; the ribs are rather more convex also, fewer in number, and rugged, as Gmelin describes it.—The colour exactly corresponds: he says it is sometimes ferruginous, with livid bands and sometimes white, with the anterior part fuscous: another variety of it is white, fasciated with a ferruginous yellowish or blueish colour.

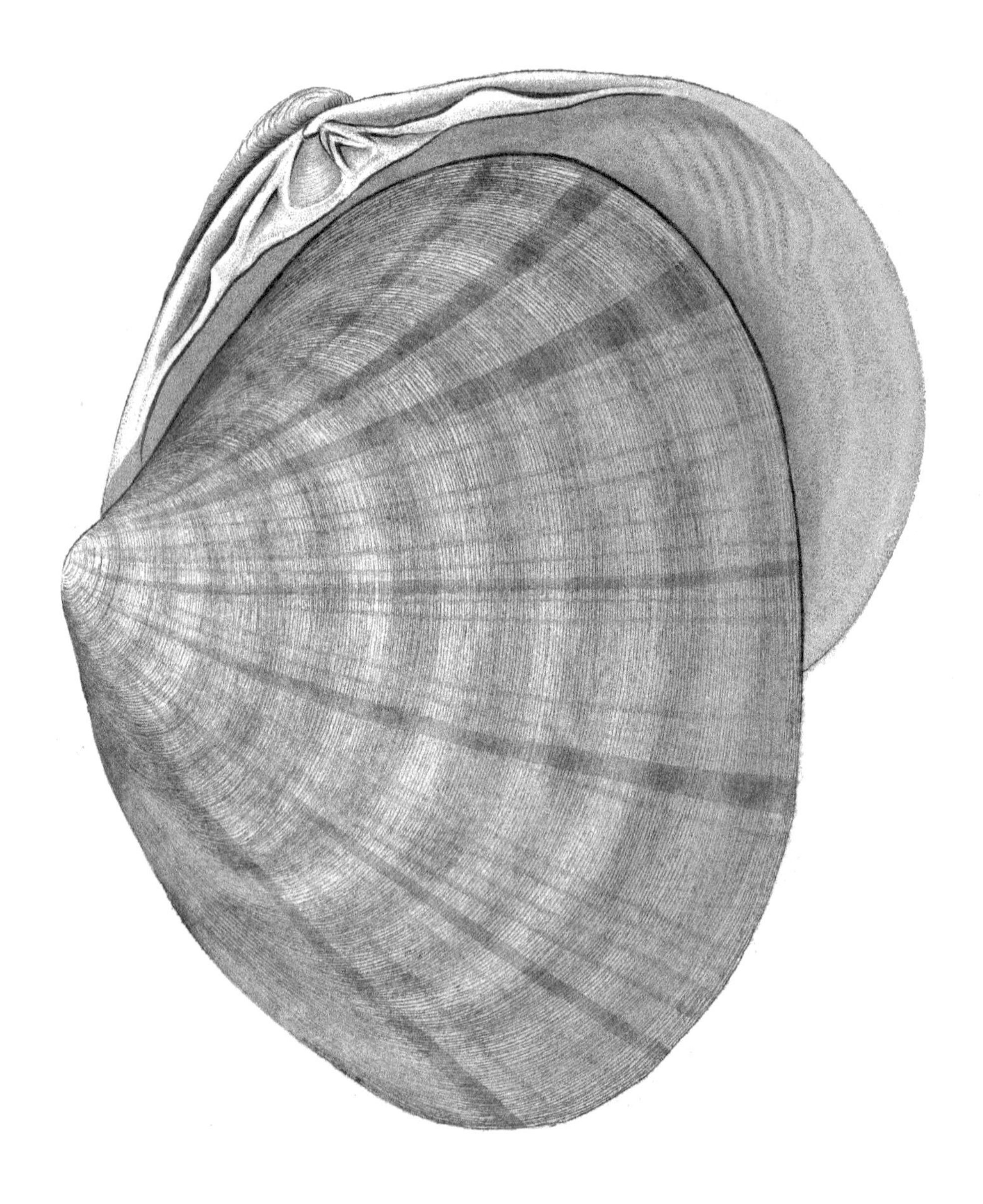

PLATE CXXV.

MACTRA GLAUCA.

GLAUCOUS MACTRA.

GENERIC CHARACTER.

Animal a Tethys, Bivalve, sides unequal. Middle tooth complicated, with a little groove on each side; lateral teeth remote.

SPECIFIC CHARACTER

AND

SYNONYMS.

Shell ovate, sordid white with brown (or glaucous) rays: very finely ſtriated transversely, and wrinkled on the anterior end.

MACTRA GLAUCA: testa ovata sordide alba glauco radiata subtilissime transversim striata anterius rugosa. *Gmel. T.* 1. *p.* 6. *p.* 3260. *ſp.* 20.

Chemn. Conch. 6. *t.* 23. *f.* 232, 233.

This is a new species as a British shell; and the conchologist is indebted to Miss Pocock for the discovery of it on our coaſt. The attention with which this lady has honoured the science has not been rewarded by this new species only: we have been favoured with several others, besides many rare kinds that have been found by her on different parts of the sea-coast, and especially on that of Cornwall, as will appear hereafter. A few shells of the species before us were met with by her in the summer of 1801, on Hale sands under Lelant in that county, and it is said by the country people they are at times found on that coaſt in some plenty.

Though hitherto unknown as a British shell, it has been before discovered in the Mediterranean sea; for there can be no hesitation in admitting it to be the shell figured by Chemnitz, as above quoted. Gmelin refers to the two figures in that Work, No. 232, and 233, for his species glauca; and the description corresponds in general with them, though not exactly in the colour of the rays.—Gmelin has another species of Mactra, *grandis*, which we at first suspected to be the same as our shell. It agrees precisely in the colour of the rays, but from the figure in Chemnitz's work, quoted by Gmelin for that species, these rays, it appears, are far more minute, than in our shell, and are also decussated by others in a concentric direction; —the outline of the latter is also different.

126

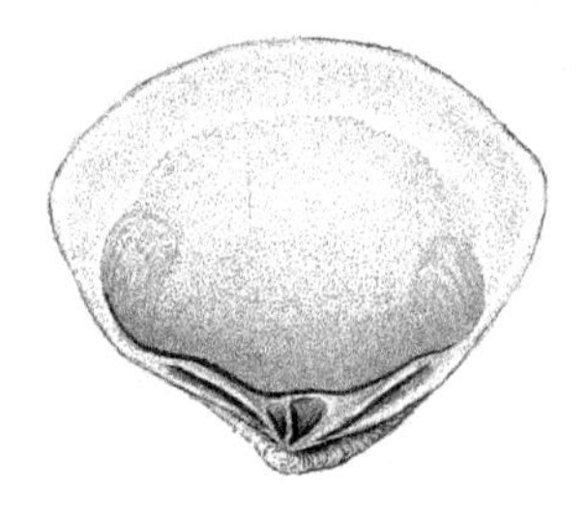

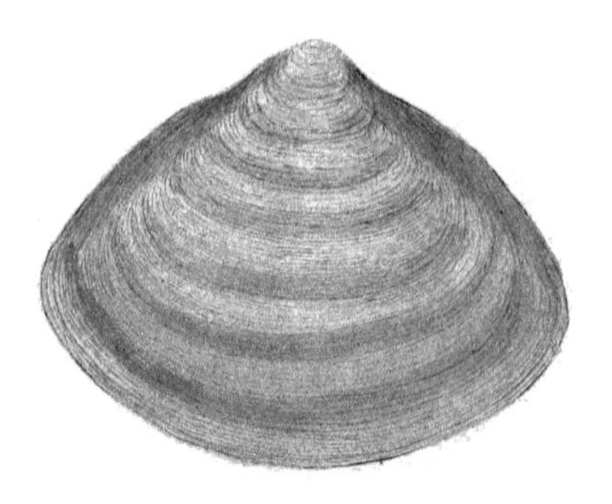

PLATE CXXVI.

MACTRA SUBTRUNCTA.

SUBTRUNCATED MACTRA.

GENERIC CHARACTER.

Animal a Tethys. Bivalve, sides unequal. Middle tooth complicated, with a little groove on each side: lateral teeth remote.

SPECIFIC CHARACTER
AND
SYNONYMS.

Shell somewhat triangular, whitish, smooth, subtruncated on each side.

MACTRA SUBTRUNCATA: testa subtriangularis, albescens, lævis lateribus subtruncatis.

Trigonella albescens lævis, lateribus subtruncatis. Subtruncata. *Da Costa. Brit. Conch. p.* 198. *No.* 34.

A thick, strong, and heavy shell, of a somewhat triangular shape, and much flattened on each side, obliquely from the beaks: the valves rather convex, the beaks pointed strong, and turned inwards.

The colour is pale or whitish, and it is externally marked with numerous fine concentric striæ.

Received from Hampshire and Devonshire; but not common. Found in a fossil state in the sand pits at Woolwich.

127

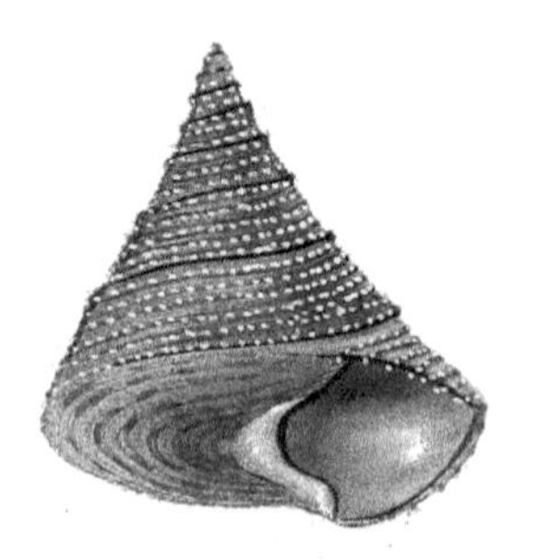

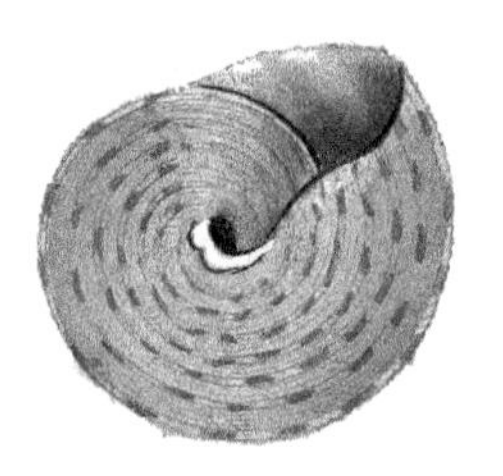

PLATE CXXVII.

TROCHUS PAPILLOSUS.

SHAGREENED TOP SHELL.

GENERIC CHARACTER.

Animal a slug. Shell conic. Aperture nearly triangular.

SPECIFIC CHARACTER.

Pyramidal, umbilicated, red; whorls with several rows of minute granulations.

TROCHUS PAPILLOSUS: pyramidalis umbilicatus ruber, series papillarum donatus. *Da Costa. Brit. Conch. p.* 38. *No.* 20.

Cul de lampe à flammes longitudinales, alternatives blanches et rouges, à stries granuleuses et umbilique: et cul de lampe marbré de blanc et de rouge, à stries circulaires granuleuses et tuberculeuses. *D' Avila, cab.* 1. *p.* 127, 128.

Da Costa acquaints us that he received "some few of these shells from Cornwall (in a great quantity of others, natives of that coast) by an intelligent gentleman of veracity and curiosity; but" adds that

E 3

writer, " must own I have never met with this species since on any other British coasts." We have since seen it among parcels of shells from the Mediterranean, and also find that it has been discovered by the late Dr. Pultney, on the North shore, Poole; and at Weymouth.

The specific character of the Gmelinian Trochus *Granatum*, seems to accord very nearly with our shell *, but it is about twice the size, and inhabits the Southern Ocean; and it is certainly more probable, as some conchologists have suggested, that it is the *variety* of Trochus Zizyphinus, described by Linnæus in the *Mus. Reginæ*, as being " tota pallida, anfractibus basi gibbis, striatis, subtitissime punctis papilloris." It differs from Zizyphinus in having the anfractus a little rounded, and the wreaths being encircled with granulated spiral ridges. The name Da Costa has already given it, is very applicable; and as it has undoubtedly escaped the notice of Gmelin, there can be no impropriety in retaining it.—It is evidently one of the rarest British species of the Trochus genus.

---

* Testa pyramidali alba coccineo varia basi subconvexa; spiræ anfractibus convexis; cingulis granorum moniformibus, primis duobus maximis. *Gmel.* 3584. *sp.* 108.

2
1
1
2

PLATE CXXVIII.

FIG. I. I.

MYTILUS EDULIS.

COMMON MUSCLE.

GENERIC CHARACTER.

The hinge toothless, and consists of a longitudinal furrow.

SPECIFIC CHARACTER
AND
SYNONYMS.

Shell smoothish, violet: valves slightly carinated in front, retuse behind: beaks pointed.

MYTILUS EDULIS: testa læviuscula violacea: valvis anterius subcarinatis posterius retusis, natibus acuminatis. *Fn. Seuc.* 2156. *Gmel. Linn. Syst. Nat. T. I.* *p.* 6. *p.* 3353. *Sp.* 11.

Mytilus vulgaris. Musculus vulgaris sublævis ex cæruleo niger. COMMON MUSCLE. *Da Costa Brit. Conch.* *p.* 216. *fp.* 48.

MYTILUS EDULIS, Edible. *Penn. Brit. Zool. T.* 4. *p.* 110. *fp.* 73.

E 4

PLATE CXXVIII.

Few species of the shell tribe are more generally diffused throughout the European and Indian seas than the Mytilus edulis; and few indeed exhibit such an infinite number of varieties, differing in size, in form, and colour; but which the critical Conchologist will yet perceive cannot with propriety be assigned to any other species.

On the Plate annexed to this description, one shell of the shape more uniformly prevalent is figured in its natural state, and another, which, having been divested of the epidermis, displays a beautiful variety of irregular purple stripes: an appearance very common in the uncoated shells, and in some sorts observable even when the epidermis is upon them. The upper and lower figures are of two shells, which, we are inclined to think, differ too widely from the common kind to be considered as a variety: the outline appears at the first sight obviously dissimilar; and the characters in general seem to mark most decidedly another species. They are both worn shells, but which we have compared with perfect specimens of Mytilus ungulatus, and apprehend there can be no doubt that they belong to that species.

Within the tropics, the common Muscle is known to attain a much larger size than in northern climates. They are found in immense beds, and adhere to other substances, or to one another, by means of a beard of a strong and silky texture, which the fish throws out. The Muscle affords a rich and palatable food; though they are not deemed wholesome by many people, who after eating them are sometimes afflicted with great swellings and convulsive motions, with eruptive blotches, shortness of breath, and even with delirium. These dreadful effects are usually attributed to some malignant poison in the little pea crab which is occasionally found in the Muscle, and may be accidentally eaten with it: others think it is in consequence of swallowing the silky byssus, or beard; and again many deem the Muscle itself poisonous. It is, however, pretty generally agreed, that they affect some constitutions more than others, and that much depends on the state of the body at the time of eating them. The disorder may be cured, or at least its malignity mitigated, by administering to the affected person a spoonful of vinegar: some recommend sweet oil, or salt and water. Da Costa observes that sudorifics, vomits, oils, &c. are the usual remedies; and the Dutch give two spoon's-full of oil, and one of lemon-juice; or, in defect of that, a little more vinegar, well shaken together, and swallowed immediately.

The Muscle is the prey of many kinds of fishes, and other creatures that inhabit the sea. On the coast of Greenland, Fabricius tells us, they are so abundant, that the dogs and ravens commonly feed on them; as do also the white game, (Ptarmigan) Eider Duck, and many others.

The seed-pearls found in the shell of the Muscle was formerly in some esteem, for medicinal purposes: these, it is well known, are the effects of a disease in the fish, analogous to the stone in the human body.

Mr. Pennant informs us, that the finest Muscles on the English coasts (where they are found in great abundance) are those called Hambleton Hookers, from a village in the county of Lancashire. They are taken out of the sea, and placed in the river Wier, within reach of the tide, where they grow very fat and delicious.

PLATE CXXVIII.

FIG. II. II.

MYTILUS UNGULATUS.

CLAWED MUSCLE.

SPECIFIC CHARACTER
AND
SYNONYMS.

Shell smooth, somewhat curved: posterior margin inflected: hinge terminal, bidentated.

MYTILUS UNGULATUS: testa lævi subcurvata: margine posteriori inflexo, cardine terminali bidentato. *Linn.—Gmel. Syst. Nat. T.* I. *p.* 6. *p.* 3354. *Sp.* 12.

Several shells of this kind were picked up by Miss Pocock, on the coast of Cornwall. It was before known as an inhabitant of the Mediterranean, but not as a British species.

Fig. II. II.—Upper and lower Figures.

129

PLATE CXXIX.

PATELLA ALBIDA.

WHITISH CHAMBERED PAP-SHELL.

GENERIC CHARACTER.

Animal a Limax: shell univalve, sub-conic without spire.

SPECIFIC CHARACTER.

Shell fragile, entire, subrotund, whitish; vertex somewhat central, and slightly pointed; lip within lateral.

PATELLA ALBIDA: testa fragilis integerrima subrotunda albida vertice subcentrali submucronato, labio interio laterali.

This rare and nondescript species of Patella was found on the coast of Cornwall, and communicated by the lady who favoured us with Mactra glauca, and Mytilus ungulatus. It differs from any of the described British shells of this genus, in having an inner lip or chamber, such as is observed in several of the exotic kinds, belonging to the first section of the genus in Gmelin's arrangement; "Labiatæ s. labio interno instructæ, testa integra."

It is an exquisitely delicate shell, and remarkably brittle: there is a specimen of this shell in the collection of William Pilkington, Esq; Whitehall.

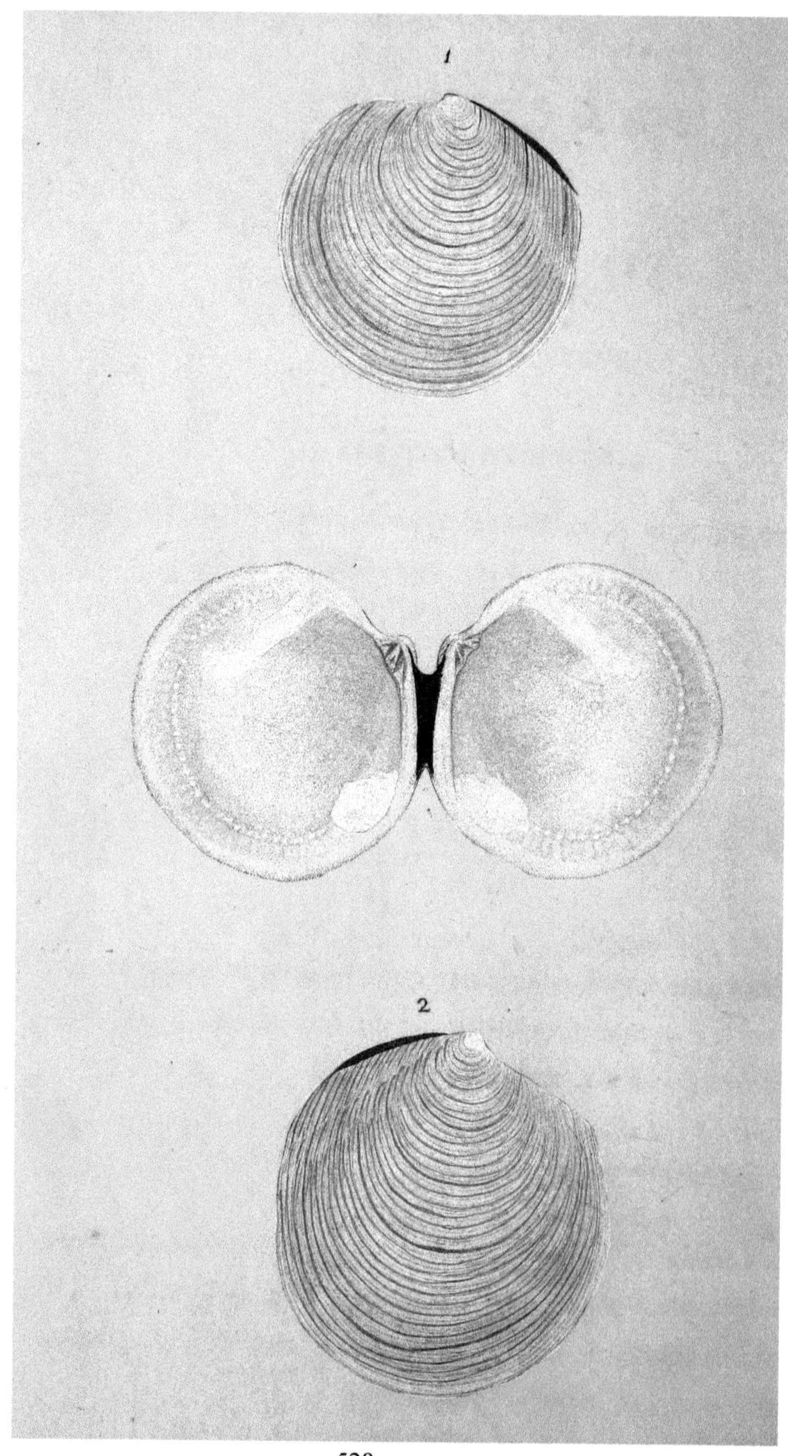
1
2

PLATE CXXX.

VENUS BOREALIS.

NORTHERN VENUS.

GENERIC CHARACTER.

Hinge furnished with three teeth; two near each other, the third divergent from the beaks.

SPECIFIC CHARACTER,

AND

SYNONYMS.

Shell lentiform, with very remote transverse, erect, membranaceous striæ.

VENUS BOREALIS: testa lentiformi: striis transversis membranaceis erectis remotissimis. *Gmel. Linn. Syst. Nat. T.* 1. *p.* 6. *p.* 3285. *sp.* 77.

This is a scarce, or very local shell on the British coasts, and seems to agree with the description of the Linnæan Venus borealis. The species was unknown to Da Costa, and is different from that which Mr. Penant describes under the same name.

PLATE CXXX.

We first discovered it on the coast of South Wales; and since that time have received it from Miss Pocock, by whom it was found on the coasts of Cornwall in some plenty.

PLATE CXXXI.

HELIX HORTENSIS.

GARDEN SNAIL.

GENERIC CHARACTER.

Aperture of the mouth contracted and lunulated.

SPECIFIC CHARACTER

AND

SYNONYMS.

Shell imperforate, globose; spotted and fasciated with brown: lip white.

HELIX HORTENSIS: testa imperforata globosa: labro albo. *Müll. Zool. Dan.—Gmel. Linn. Syst. Nat. T.* I. *p.* 6. *p.* 3649. *sp.* 109.

Cochlea vulgaris fusca, maculata & fasciata. VULGARIS. *Da Costa, Brit. Conch. p.* 72. *sp.* 39.

Helix Hortensis. Garden Snail. *Penn. Brit. Zool. No.* 129. *tab.* 84. *fig.* 129.

HELIX LUCORUM. *Linn.*

The Common Garden Snail is, we think, without doubt, the Helix lucorum of Linnæus, and most other writers. Gmelin calls it hortensis, and defines the specific character of lucorum to be "testa imperforata subrotunda lævi fasciata: apertura oblongo fusca." It more generally inhabits the southern parts of Europe; is larger, and whiter than hortensis; and the lip is brown.

This is a most variable species in its colours and markings, and it may still be doubted whether hortensis and lucorum be perfectly distinct. They have certainly been confounded by almost every Conchologist.

On the manners of a creature which is so generally known, as the Common Garden Snail, it is surely needless to enlarge: its mode of courtship is, however, so curiously related, that it should not entirely escape remark; and were it not attested by writers of the first authority, with the reader, we might be guilty of no small degree of scepticism, as to believing it. Each of these animals, it seems, are furnished, at a certain season, with a number of little pointed darts, which are contained within a cavity on the right side of the neck. When the Snails approach within two or three inches of each other, a scene of hostility is observed to commence: each discharging at its antagonist these darts, with considerable force, at the other; this battle continues till the reservoir be exhausted of these offensive weapons, and then a perfect reconciliation takes place between them. The eggs are about the size of peas, and perfectly round.

Snails are used with success in some consumptive cases, and an excellent cement, to fasten china, may be made of the saliva, or humours, mixed with quick lime and white of eggs, according to

Lister, &c. It feeds on all kinds of vegetables and fruits, and is consequently very destructive in orchards and gardens. Snails couple about May or June.

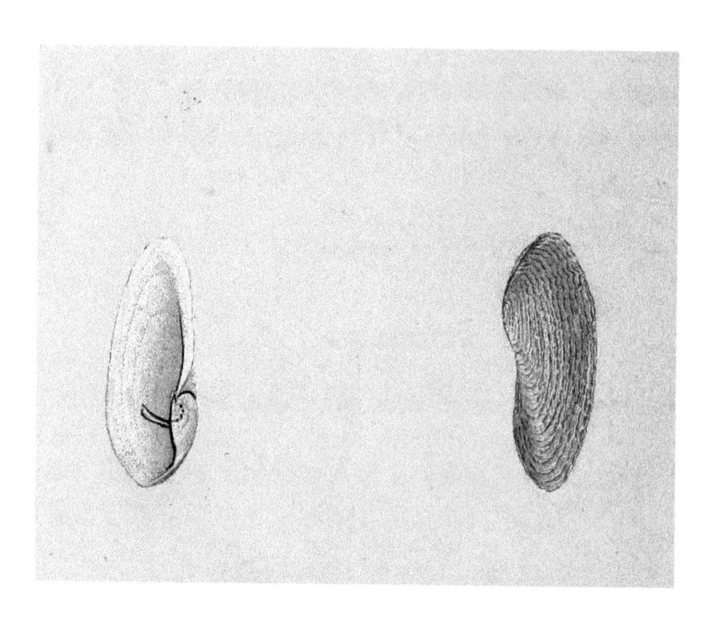

PLATE CXXXII.

PHOLAS CANDIDA.

WHITE PIDDOCK.

GENERIC CHARACTER.

Animal Ascidia. Shell bivalve, opening wide at each end, with several lesser valves at the hinge. The hinges folded back and connected by a cartilage.

SPECIFIC CHARACTER,

AND

SYNONYMS.

Shell oblong, muricated with striæ crossing each other.

PHOLAS CANDIDA: testa oblonga undique striis decussatis muricata. *Mus. Lud. Ulr.* 469. *n.* 7.—*Gmel. Linn. Syst. Nat. T.* 1. *p.* 6. *p.* 3215. *sp.* 4.

Pholas tenuis candidus ovatus decussatim striatus. Candidus. *Da Costa. Brit. Conch. p.* 246. *sp.* 66.

Concha candida, dupliciter striata et veluti aculeata. *List. H. An. Angl. p.* 193. *tit.* 39. *tab.* 5. *fig.* 39.—Pholas alter. *App. H. An. Angl. in Goed. p.* 37. *tab.* 2. *fig.* 4 *and* 6.—Pholas parvus asper. *H. Conch. tab.* 435. *fig.* 278.

Pholas latus; short Pierce-stone or Pholade. *Petiv. Gaz. tab.* 79. *fig.* 11.

Pholas candidus. *Penn. Brit. Zool. T.* 4. *No.* 11. *tab.* 39. *fig.* 11.

PLATE CXXXII.

The Pholas candida is rather a scarce species, and differs from P. Dactylus (hians of Dr. Solander) in not being above one fourth of its size: of a more oval shape, and having both ends equally rounded; in other respects it resembles it. Da Costa seems to entertain some doubt whether it may not be really a variety in growth, than a distinct species from Dactylus. It is found on the same coasts as the latter.

133

PLATE CXXXIII.

SABELLA TUBIFORMIS.

STRAIGHT TUBE SABELLA, OR SAND SHELL.

GENERIC CHARACTER.

Animal a Nereis, with a ringent mouth, and two thicker tentacula behind the head. Shell tubular, and composed chiefly of sand, agglutinated to a membranaceous tube.

SPECIFIC CHARACTER

AND

SYNONYMS.

Shell solitary, simple, tube-shaped, straight; gradually tapering; and composed of brownish sand.

SABELLA TUBIFORMIS: testa solitaria simplici tubiformi recta sensim attenuata: granis arenaceis fuscis.

SABELLA GRANULATA. *Linn.* 1268 ?—*Martini*, 4. *t.* 4. 28 ?

SABELLA TUBIFORMIS. TUBE SABELLA. *Penn. Brit. Zool.* 4. *sp.* 163.

STRAIGHT SABELLA. *Pult. Hist. Dorset.*

The Sabella tubiformis of Pennant is believed by some conchologists to be the same species as Linnæus names *granulata*, but this must certainly admit of doubt: the latter, from the description, appears to be slightly incurvated, whilst Pennant's shell is perfectly straight; a difference, perhaps, sufficient to constitute two distinct species. *Martini* figures a kind of Sabella, which he considers as the S. granulata of Linnæus, and in that figure the curvature at the narrowest end is very apparent: with Martini we consider that as the true Sabella granulata of Linnæus, and not having observed the same character in any of those shells that have occurred to notice on our own coasts, have thought it most adviseable to retain the name Mr. Pennant had previously given it.

Gmelin seems to be under some doubts respecting this shell, for he entirely omits the Linnæan *granulata* among the species of this genus.—Sabella Belgica of Gmelin, which some have conjectured to be the *S. tubiformis* of Pennant, is undoubtedly different, according to *Klein* and *Martini*.

Our shell is of an elegant form and remarkably delicate: it consists of two coatings, the inner one of which is composed entirely of grains of sand, and the outer one of sand intermixed with little fragments of shells. This kind is local: it is supposed to inhabit deep waters only, and is sometimes found upon the shore after a high sea.

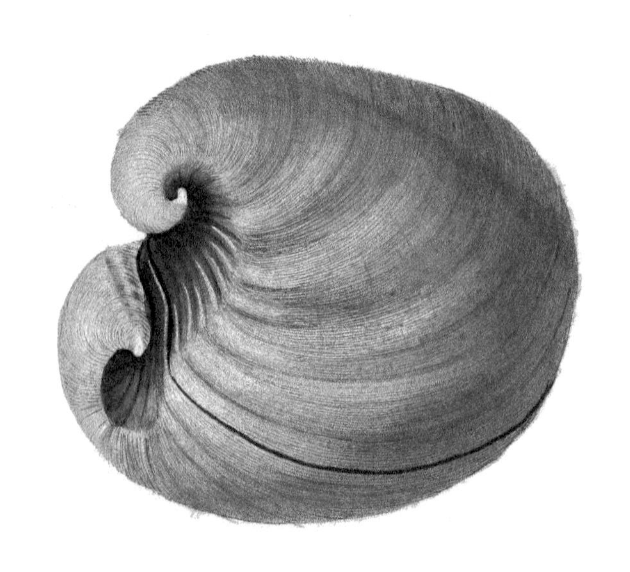

PLATE CXXXIV.

CHAMA COR.

HEART-SHAPED CHAMA.

GENERIC CHARACTER.

Animal a Tethys. Shell bivalve, and rather coarse. Hinge with a callous gibbosity, obliquely inserted into an oblique hollow : anterior slope closed.

SPECIFIC CHARACTER

AND

SYNONYMS.

Shell subrotund, and smooth: beaks recurved: anterior slope gaping.

CHAMA COR: testa subrotunda lævi: natibus recurvatis rima hiante. *Mus. Lud. Ulr.* 516. *n.* 88.—*Gmel. Syst. Nat. T.* 1. *p.* 6. *p.* 3299. *sp.* 1.

As an inhabitant of the Mediterranean, the Adriatic, and Caspian sea, this curious shell has been long since known; but as a British species has not been before described. It was discovered several years ago in the seas about the islands in the North of Scotland, by Mr. Agneu, Gardener to the late Duchess of Portland; and since that time, a specimen of it was dredged up, by T. Swainson, Esq. of the Custom-house, London.

It is a delicate shell, and represented in the annexed plate of the natural size. By some it is called the Bull's-heart Cockle, but very improperly, because it cannot be considered in any manner of the Cardium genus.

135

PLATE CXXXV.

ARCA LACTEA.

MILKY-WHITE ARK.

GENERIC CHARACTER.

Bivalve, valves equal. Teeth of the hinge numerous, and inserted between each other.

SPECIFIC CHARACTER AND SYNONYMS.

Shell somewhat rhombic, with obsolete, decussated striæ, and diaphanous.

ARCA LACTEA: testa subrhomboidea obsolete decussatim striata diaphana. *Gmel. Linn. Syst. Nat. T.* 1. *p.* 6. *p.* 3309. *sp.* 15.

ARCA LACTEA: testa subrhomboidea obsolete decussatim striata diaphana, natibus recurvis, margine crenulato. *Linn. Syst. Nat. p.* 1141. *No.* 173.

Pectunculus exiguus albus, admodum tenuiter striatas. *List. H. Conch. tab.* 235. *fig.* 69.

Mytulus Garnseiæ albus, parvus tenuiter cancellatus. *Petiv. Gaz. tab.* 73. *fig.* 1.

G 3

PLATE CXXXV.

Arca. Parva alba cancellata. Lactea. *Da Costa Brit. Conch. p.* 171. *sp.* 14.

ARCA CRINITA. *Soland. Mus. Portland.*
Pultney. Nat. Hist. Dorsets.

There does certainly exist among conchologists a strange discordance of opinion, respecting the species of Arca before us: some believing it to be the very shell which Linnæus described under the name of *lactea*, and others maintaining the contrary. Dr. Solander, and Dr. Pultney, were persuaded that Linnæus intended a shell in some particulars not unlike this, but specifically different: the same shell in fact which Lister figures No. 67, in his Conchology. The opinions of two such respectable naturalists deserve every consideration, but however we may be disposed to abide by their authority in some instances, we cannot, for the following reasons, assent to it in the present.

Dr. Pultney premises his description by naming this species Arca crinita, a name which it appears Dr. Solander had previously given to it, in his catalogue of the Portland Museum: he observes very justly, that it is the *A. barbata* of Pennant, but not the barbata of Linnæus: refers to Da Costa's figure, Plate 11, fig. 5; and calls it in English the HAIRY ARK-SHELL. His description is in these words, " A small oblong subrhomboidal shell: in its recent state covered with a brown epidermis. Underneath the shell is minutely striated in the longitudinal direction; commonly not much larger than a horse-bean. I found one on the North shore at Poole, more than half an inch long, and seven-eights wide. It is said to be more frequent on the shores of Guernsey island; but is very rare on our coasts. This shell has commonly been described as the *A. lactea* of Linnæus; but that shell is much larger, and is strongly striated in a reticulated manner."

In the first place, this does not so clearly agree with the Linnæan description as might be expected. In the early editions of the *Systema Naturæ*, Linnæus expressly describes his shell as being of a sub-rhombic form, with obsolete decussating striæ, diaphanous, with re-curved beaks, and crenulated margin; and the only alteration in the specific character to the last edition of that work, by *Gmelin*, is, that the beaks and margins are not mentioned. It hence is evident that Dr. Pultney is mistaken in the Linnæan *lactea*, when he says it is *strongly striated* in a reticulated manner; for Linnæus exactly de-scribes it as having the reticulating striæ *obsolete*; and notwith-standing that the longitudinal striæ in the shell found on our coast are rather prominent in some specimens, they far more frequently appear altogether obsolete, and are decussated by fine transverse striæ, precisely as Linnæus mentions. There is another circumstance equally remarkable, Linnæus exactly tells us his *lactea* is the size of a horse-bean, and Dr. Pultney, after observing that his British shell *crinita* "are commonly not much larger than a horse-bean," adds, that the *lactea* of Linnæus is "much larger."

That there may have been some erroneous references among authors to the two shells figured by Lister, No. 67 and 69, of which we are not aware, is not impossible, but *Gmelin* refers to *List. Conch.* 69. *A. T. Guernsey*, for the Linnæan *lactea*, and that shell clearly corresponds both with our specimen, and the shell described by Linnæus, pectunculus exiguus albus admodum tenuiter striatis. *List.—Petiver* is equally satisfactory, Mytilus Garnseiæ

G 4

albus, parvus tenuiter cancellata. Da Costa is not so very clear, nor is his figure expressive, because it represents the transverse ridges too prominent. Chemnitz gives a larger specimen of it than any we have seen; and notwithstanding his references to Da Costa and others, may not be of the same species.——The shell figured by Lister, No. 67, which seems to have occasioned the confusion before alluded to, it should be observed, is described by Gmelin, in these words: testa subrhomboidea decussatim striata alba: natibus approximatis, vulva cordata, *Arca reticulata.*

Hence it is presumed the shell found on our coast ought still to be considered as the *Arca lactea* of Linnæus, and that the shell which is strongly striated in a reticulated manner, and so much larger; and has been hitherto regarded by some as that species, was never described, or even noticed by Linnæus. Nothing can be more evident than that this is the opinion of his editor Gmelin, both from his references to the figures of lactea, and by his giving a new name and character to that very shell in Lister which has been mistaken for it, and which he conceived Linnæus had omitted.

This kind is generally considered as an inhabitant of Dorsetshire, Devonshire, and Cornwall.

PLATE CXXXVI.

HELIX ARBUSTORUM.

SINGLE-STREAK SNAIL.

GENERIC CHARACTER.

Aperture of the mouth contracted and lunulated.

SPECIFIC CHARACTER.

Shell umbilicated, convex, aperture somewhat orbicular, lip reflected, with a single dark spiral band.

HELIX ARBUSTORUM: testa umbilicata convexa acuminata, apertura suborbiculari bimarginata: antice elongata. *Linn. Syst. Nat. p.* 1245. *No.* 680.

Cochlea testa utrinque convexa subcinerea: fascia solitaria grisea, labro reflexo. *Linn. Fn. Suec.* 1. *p.* 370, *No.* 1295. II *No.* 2184.

Cochlea maculata, unica fascia pulla, angustioreque, per medium anfractus, insignita. *List. H. An. Angl. p.* 119. *tit.* 4. *tab.* 2. *fig.* 4.

Cochlea subflava maculata atque unica fascia castanei coloris per medium anfractus, insignita. *Phil. Trans. No.* 105. *fig.* 4.

PLATE CXXXVI.

Cochlea subumbilicata, maculata, unica fascia angusta, per medium anfractus insignita. UNIFASCIATA, single streak, *Da Costa. Brit. Conch. p.* 75. *sp.* 40.

Helis arbustorum. *Shrub. Penn. Brit. Zool. No.* 130. *tab.* 85. *fig.* 130.

Like the two species of Helices, nemoralis and hortensis, this shell varies exceedingly in its colours and markings in different shells; but the single narrow spiral band very uniformly distinguishes it from any other. According to Lister and Petiver, it is found in shady hedges, and is frequent in rushy, moist, or marshy meadows. Da Costa says, it is also found in coppices and shrubberies; but, from his own observations, it is not a common kind: we have seen it in woods, though not often.

It may deserve remark, that this shell is generally found empty. Muller accounts for this, by acquainting us, that the animal is the prey of young Newts. Sometimes when the animal is living, the shell is of a light chesnut colour, finely marbled, mottled, or otherwise variegated, with a deep chocolate coloured brown.

PLATE CXXXVII.

OSTREA JACOBÆA.

PILGRIM OYSTER, OR SCALLOP.

GENERIC CHARACTER.

Animal a Tethys. Shell bivalve, with the valves mostly unequal and slightly eared. Hinge without teeth, but furnished with an ovate hollow, and in general lateral transverse furrows.

SPECIFIC CHARACTER
AND
SYNONYMS.

Shells with fourteen angular and longitudinally striated rays.

OSTREA JACOBÆA: testæ radiis quatuordecim angulatis longitudinaliter striatis. *Linn.—Gmel. Syst. Nat. p.* 3316. *sp.* 2.

Pecten magnus inæquivalvis operculo ſubrufo, fornix vero albus costis angulatis et canaliculatis. *Jacobæus. Da Costa Brit. Conch. p.* 143. *sp.* 2.

P. JACOBÆUS Lesser. *Penn. Brit. Zool. No.* 62. *tab.* 60. *fig.* 62.

This large and handsome kind of Scallop is rare on the English coasts: in the Mediterranean it is more frequent, and from that

that circumstance, is supposed to be the same species as was formerly worn by pilgrims, who visited the holy-land. Da Costa received it from Scarborough in Yorkshire, and also from the coasts of Cornwall and Dorsetshire. Dr. Pultney remarks, that it is rare on the Dorsetshire coast, but has been fished up at Weymouth and at Poole.

The upper valve is of a reddish brown, flat, and rather depressed at the beak: the depression is generally white, and mottled, or otherwise marked with brown, or dusky red; and often with prettily disposed marks, resembling the heads of arrows. The under valve is usually white, or faintly tinged with brown, and has the rays remarkably prominent and angular; a character by which it may be easily distinguished from the Common Scallop, *Ostrea maxima*, which greatly resembles it, but in which the rays are rounded.

PLATE CXXXVIII.

VOLUTA TRIPLICATA.

THREE PLAITED VOLUTE.

GENERIC CHARACTER.

Animal Limax. Shell spiral, narrow, without a beak: columella or pillar, twisted or plaited.

SPECIFIC CHARACTER.

Shell ovate, smooth and brown: spire rather pointed: whorls six: pillar with three plaits.

VOLUTA TRIPLICATA: testa ovata lævi brunnea, spira acutiuscula: anfractibus sex, columella triplicata.

Turbo sex anfractibus apertura ovali bidentata. *Walker. Test. min. rar. fig.* 50?

This curious little Volute has not we believe been either figured or described by any Author. In its general appearance it approaches pretty nearly to the shell figured by Walker, as above quoted; but if the same, that writer has certainly described it with no very great degree of accuracy; for he speaks only of two plaits, or teeth in the aperture, and calls it the double-toothed Turbo ; although from its plaited pillar, it is evidently a Volute : and the third plait, which with the two others, characterize the species, is obviously too large to have been overlooked. Da Costa, we are persuaded, thought they were both the same, for in a *MS. note* attached to this shell in his collection, he says, " This shell is figured by Walker," and certainly no figure in the volume of that Author corresponds with it, except that to which we refer above.

Walker speaks of his shell as being very common on the roots of rushes, in marshes, near Faversham : The *habitat* of our shell we have not been able to ascertain, having never found it in a living state.

PLATE CXXXIX.

SABELLA ALVEOLATA.

HONEY-COMB SABELLA, OR SAND SHELL.

GENERIC CHARACTER.

Animal a Nereis, with a ringent mouth, and two thicker tentaculæ behind the head. Shell tubular, and composed chiefly of sand, agglutinated to a membranaceous tube.

SPECIFIC CHARACTER
AND
SYNONYMS.

Composite, consisting of numerous parallel tubes, with somewhat funnel-shaped aperture.

SABELLA ALVEOLATA: testa composita concamerationibus numerosis: aperturis sub-infundibuliformibus.

SABELLA ALVEOLATA: testa composita concamerationibus numerosis poro communicantibus. *Gmel. Linn. Syst. Nat. T.* 1. *p.* 6. *p.* 3749. *Sp.* 3.

SABELLA ALVEOLATA, HONEY-COMB. *Penn. Brit. Zool. Vol.* 4. *p.* 147. *sp.* 162.—*Ellis Coral. p.* 90. *t.* 36.

PLATE CXXXIX.

We have constantly observed this kind of Sabella to be an inhabitant only of low rocky shores, that lie within reach of the tide at high water; and although it is composed entirely of broken shells and sea sand; and those apparently very slightly agglutinated by the animals that form and inhabit it, in its native element it acquires a considerable comparative degree of tenacity from the saline particles with which it is impregnated, and is capable of resisting the reiterated dashing of the waves without material injury.

In the more sheltered crevices of the rocks, where the animal has ample scope to expand its sandy dwelling secure from mutilation, the upper surface of a mass of these shells has a very elegant appearance; the whole being beautifully foliated with the single or bipartite funnnel-shaped lip, which each animal forms at the opening of his cell. This appearance is represented in the plate subjoined, and is a circumstance the more deserving attention, because it has entirely escaped the notice of Mr. Pennant, and perhaps of every other Naturalist, except Mr. Ellis; who in his History of Corals, figures it with a foliated surface, under the title of Tubularia arenosa anglica, from a specimen brought from Dieppe. The latter does not however agree precisely in figure with any that have occurred to our notice. Mr. Pennant evidently represents a poor mutilated fragment, in which the openings of the cells are shewn like so many rounded perforations, in a somewhat uneven surface, and is just as it appears in masses on the sea-beach, that have been trodden under-foot; or pieces that have been much worn, and thrown loose upon the shore.

These shells are three or four inches in length, and where they are found are generally abundant, but they seem to be very local. Mr. Pennant says it is found on the Western coasts of Anglesea, near Cricceth in Caernarvonshire, and near Yarmouth. We have seen it on other parts of the sea coast of North Wales, and in equal plenty on those of South Wales likewise. On the coast of Dorsetshire, according to Dr. Pultney, fragments are found very frequently, but none very perfect. In the History of that County the following account is given of it.—" GREGARIOUS SABELLA. This is composed of a mass of fine sand, and particles of broken and finely comminuted shells, aggregated by vermiform animals of the Nereis genus, each lodged in its separate tube close to, but not interfering with each other. All the tubes end in orifices on the upper or the same surface. I but once saw a mass of this kind, about the size of a large apple, on the beach, a mile east of Weymouth ; but I suspect it is not very uncommon, as fragments are very frequent."

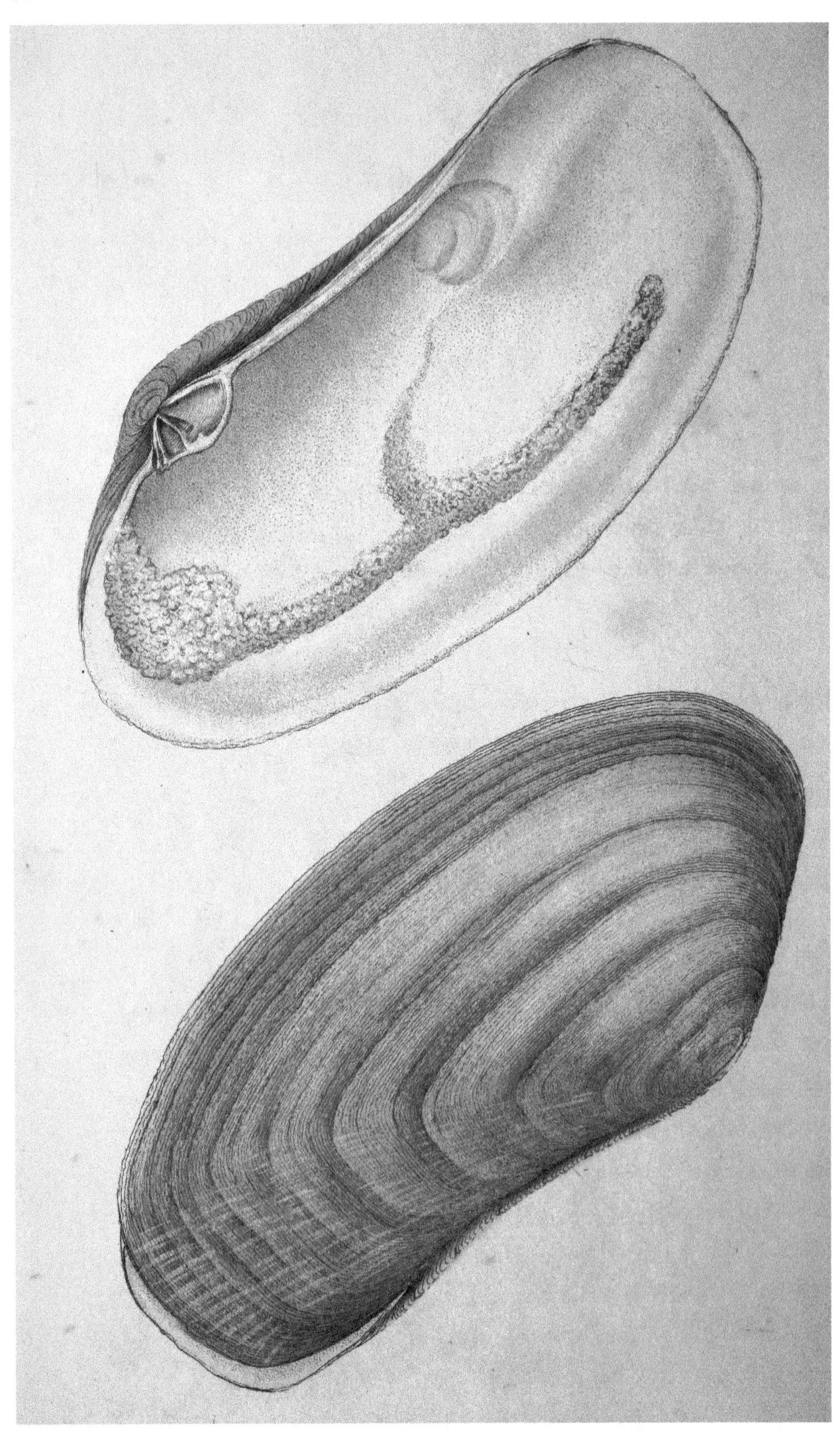

PLATE CXL.

MACTRA HIANS.

GAPING, OR OBLONG MACTRA.

GENERIC CHARACTER.

Animal a Tethys. Shell bivalve: valves equal; sides unequal: middle tooth of the hinge complicated, with a small hollow: lateral teeth remote and inserted into each other.

SPECIFIC CHARACTER.

Shell oblong, rather arcuated, coarse, gaping anteriorly, and the hinge placed very far back.

MACTRA HIANS: testa oblonga sub-arcuata rudi anticè hiante, cardine subterminali.

MACTRA HIANS. *Soland. Mus. Port.*

Da Costa and some other Conchologists have very erroneously been led to conclude, that this, and Mactra lutraria, are the same species, although they differ in almost every particular. The present shell is much wider in proportion to the length than *M. lutraria*; and of a more incurvated shape: it is also a thicker and coarser shell; and has the hinge placed much further back than in the other species. Dr. Solander, who described it in the catalogue of the Portland Museum, gave it the specific name of hians, from its re-

H 2

markable gaping at the anterior end, and a name so applicable we thought it best to retain.

The cicatrix of the animal, in this kind, is different from that of M. lutraria, a circumstance that has not escaped the observation of Dr. Pultney. This author tells us, the Mactra hians is thrown up in considerable quantities on all the smooth beaches he has seen on the coast of Dorsetshire, particularly on the North shore at Poole, opposite Branksea isle ; and that he has also seen it on the beaches at Studland, Swanage, and Weymouth. We have observed it, but not in abundance, on other sea-coasts.

PLATE CXLI.

MYTILUS RUGOSUS.

RUGGED MUSCLE.

GENERIC CHARACTER.

The hinge toothless, and consists of a longitudinal furrow.

SPECIFIC CHARACTER
AND
SYNONYMS.

Shell rhombic oval, rugged, obtuse at the ends and antiquated.

MYTILUS RUGOSUS: testa rhomboideo-ovali rugosa obtusa antiquata. *Linn. Syst. Nat. p.* 1156. *No.* 249.

Pholas noster, sive concha intra lapidem quendam cretaceum degens. *List. H. An. Angl. p.* 1722. *tit.* 21. *tab.* 4. *fig.* 21.

Mytilus parvus rhomboidea-ovalis, subalbescens, rugosus. *Rugosus*, Rugged, *Da Costa Brit. Conch. p.* 223. *sp.* 52.

The Mytilus rugosus is rather a rare species on our coasts. Lister notes it from Hartlepool in Durham, and Da Costa says on the coast of Yorkshire, about Scarborough, Whitby, &c. It is found in incredible abundance, niched or burrowed, in the rocks of lime-stone, &c,

H 3

PLATE CXLI.

In habit and manners of life this species greatly resembles the Pholades, each forming for itself a separate apartment within the hard clay, or solid stone: this it pierces when young, and afterwards continues to enlarge the cell as it increases in bulk, without widening the aperture; so that when full grown, the shell cannot easily be taken whole out of the cell, without breaking the stone in which it is contained.

This shell may perhaps be arranged with equal propriety with the *Myæ* as the *Mytili*, notwithstanding that it is admitted among the latter by most collectors.

PLATE CXLII.

MYA GLYCYMERIS.

LARGE MYA.

GENERIC CHARACTER.

Animal Ascidia. Shell bivalve, gaping in general at one extremity: hinge with a thick patulous tooth; seldom more than one, and that not inserted into the opposite valve.

SPECIFIC CHARACTER
AND
SYNONYMS.

Shell gaping at both ends: very thick, lamellous oblong-oval, with transverse rugose striæ: first tooth of the hinge very thick.

MYA GLYCYMERIS: testa utrinque hiante crassissima lamellata oblongo ovata transverse striato rugosa, cardinis dente primario crassissimo. *Gmel. Syst. Nat. p.* 3222. *sp.* 17.

MYA GLYCYMERIS: testa sub-ovata oblonga, ponderosa, ventricosa, utrinque hians, antice et postice quasi oblique truncata, dente cardinali crassissimo. *Chemn. T.* 6. *p.* 33.

Telline beante, *Favart D'Herbigny. Dict. T.* 3. *p.* 358.
List. n. Conch. 6. *t.* 3. *f.* 25. *Born. t.* 1. *f.* 8.

H 4

PLATE CXLII.

A species of Mya admitted with some doubt as a British shell: it is a kind acknowledged however as such, by collectors of English Natural History in general; and is said to have been undoubtedly fished up in the deep waters between the Dogger-Bank and the eastern coast of England.

The Mya Glycymeris is the largest of its genus, and is an inhabitant of most parts of the European sea. In the Mediterranean, and on the northern coasts of Spain, it is not uncommon: on the coast of France it is also found sometimes.

PLATE CXLIII.

BULLA PATULA.

PATULOUS BULLA.

GENERIC CHARACTER.

Animal Limax. Shell rather convoluted at one end, suboval: aperture oblong.

SPECIFIC CHARACTER
AND
SYNONYMS.

Shell ovate, smooth, and somewhat beaked at both ends; that at the base produced and sub-umbilicated: lip entire.

BULLA PATULA: testa ovata lævi sub-birostri: basi productiori sub-umbilicata, labro integro.

BULLA PATULA. Open. B. with one one end much produced and fuciform. The aperture very patulous. *Penn. Brit. Zool. V.* 4. *p.* 117. *sp.* 85.

Pennant seems to be the only author who has noticed this species. Da Costa imagined, from the description given by that author, that it was of the same kind as that which he inserted in his British

PLATE CXLIII.

Conchology, under the name of Bulla, the Bubble (Aperta, Linn.) and refers to the figure in the work of Pennant accordingly; but in this instance he was much mistaken, for the two shells are perfectly distinct; and it appears certain, that the shell described by Da Costa was as much unknown to Pennant, as that of Pennant was to Da Costa. The shell of the latter is figured in Plate 120 of this work, and the true Bulla patula of the other is figured in the Plate annexed.

This we apprehend to be one of the rarest of the British shells hitherto discovered; Pennant notes it from Weymouth, and refers for his specimen to the Portland Cabinet: our shell is from Weymouth likewise.

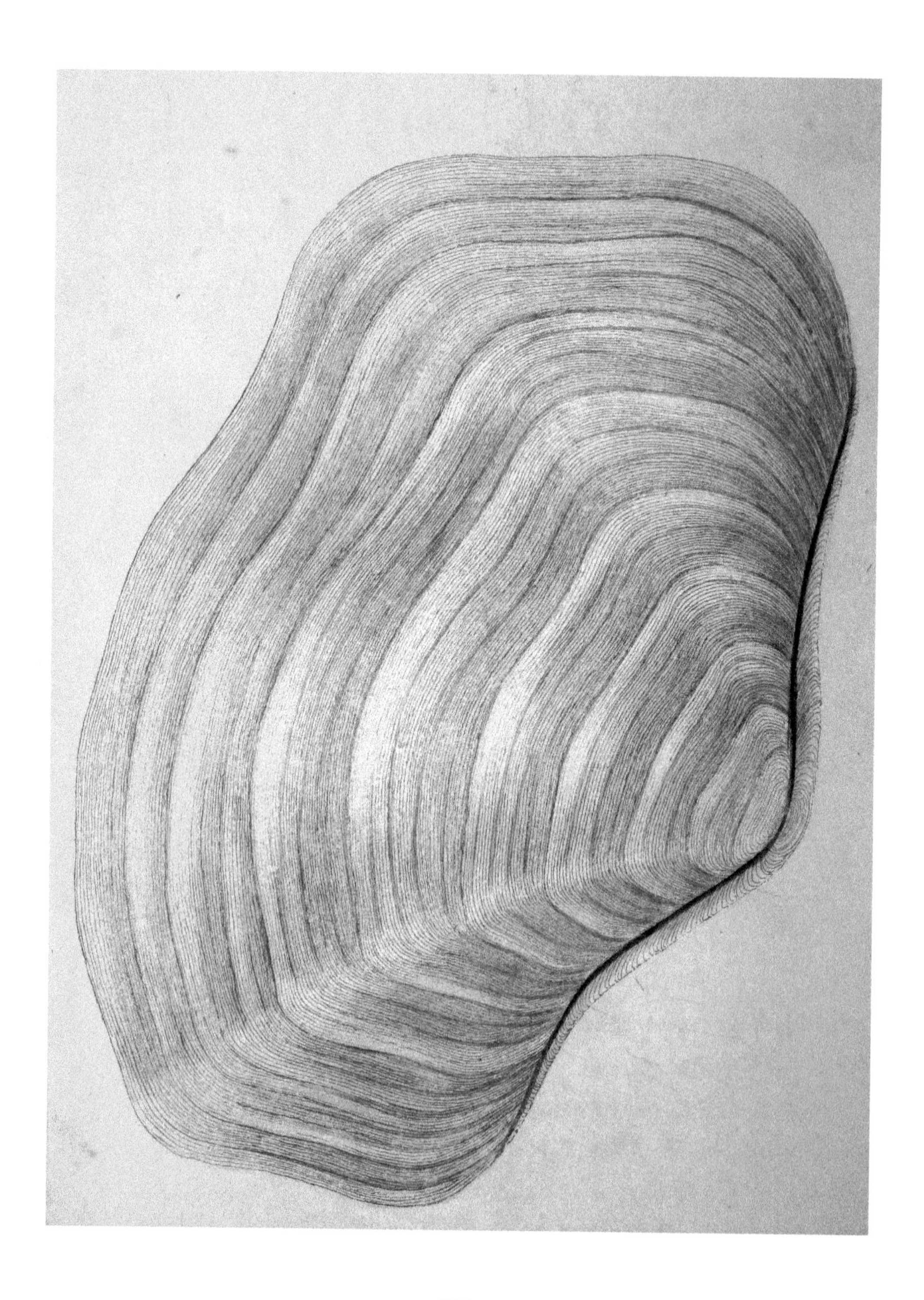

PLATE CXLIV.

NERITA NITIDA.

GLOSSY NERIT.

GENERIC CHARACTER.

Animal Limax. Shell univalve, spiral, gibbous, and rather flat beneath: aperture semi-orbicular, or semi-lunar: pillar lip transverse truncated and flattish.

SPECIFIC CHARACTER.

Shell smooth, white, and glossy: spire rather pointed: umbilicus half closed.

NERITA NITIDA testa lævi nivea nitida: spira sub-mucronata umbilico semi-clauso.

Among the reserved shells intended by Da Costa for a second edition of his Conchology, we find two specimens of this species of Nerita, with a *MS.* memorandum, importing that he had received one of them from Mr. Church, and that the other was in his possession before. On this vague authority, we did not think it incumbent to insert the shell in the present Work, especially since its habitat was not mentioned; but the same kind was discovered, in the course of last summer, upon the coast of Scotland, near Caithness, by A. Macleay, Esq. and we can no longer hesitate to insert it as an undoubted British species.

In the annexed plate it is represented of the natural size. It is a remarkable little shell, and is not to the best of our knowledge, mentioned in the work of any author.

INDEX

VOL. IV.

LINNÆAN ARRANGEMENT.

* MULTIVALVIA.

BIVALVIA. CONCHÆ.

INDEX.

INDEX TO VOL. IV.

ACCORDING TO

HISTORIA NATURALIS TESTACEORUM BRITANNIÆ OF DA COSTA.

PART I.

UNIVALVA NON TURBINATA.

GENUS I. PATELLA. LIMPET, FLITHER, OR PAP SHELL.

PART II.

UNIVALVIA INVOLUTA.

GENUS 5. BULLA. DIPPER.

I N D E X.

PART III.

UNIVALVIA TURBINATA.

GENUS 7. TROCHUS. THE TOP.

TERRESTRES. LAND.

* MARINÆ. SEA.

GENUS 10. COCHLEA SNAILS.

* TERRESTRES. LAND.

GENUS 12. STROMBIFORMIS. NEEDLE SNAIL.

* MARINÆ. SEA.

ORDER 2.

BIVALVES.

GENUS 1. PECTEN ESCALLOP.

INDEX.

GENUS 5. ARCA, ARKS, OR BOATS.

* MARINÆ. SEA.

GENUS 6. CARDIUM. HEART COCKLE.

* MARINÆ. SEA.

GENUS 11. MYTILUS MUSCLE.

* FLUVIATILES. RIVER.

* MARINÆ SEA.

PART III.

GENUS 13. CHAMA GAPERS.

* MARINÆ. SEA.

INDEX.

GENUS 14. SOLEN. SHEATH OR RAZOR SHELL.

PART IV.

MULTIVALVES.

GENUS 16. PHOLAS PIDDOCKS.

ALPHABETICAL INDEX TO VOL. IV.

INDEX.

END OF VOL. IV.

Printed by Bye and Law, St. John's Square, Clerkenwell.

THE

NATURAL HISTORY

OF

BRITISH SHELLS,

INCLUDING

FIGURES AND DESCRIPTIONS

OF ALL THE

SPECIES HITHERTO DISCOVERED IN GREAT BRITAIN,

SYSTEMATICALLY ARRANGED

IN THE LINNEAN MANNER,

WITH

SCIENTIFIC AND GENERAL OBSERVATIONS ON EACH.

VOL. V.

BY E. DONOVAN, F.L.S.

AUTHOR OF THE NATURAL HISTORIES OF

BRITISH BIRDS, INSECTS, &c. &c.

LONDON:

PRINTED FOR THE AUTHOR,

AND FOR

F. AND C. RIVINGTON, N° 62, ST. PAUL'S CHURCH-YARD;

BY BYE AND LAW, ST. JOHN'S SQUARE, CLERKENWELL.

1803.

THE

NATURAL HISTORY

OF

BRITISH SHELLS.

PLATE CXLV.

TEREDO NAVALIS.

SHIP WORM.

GENERIC CHARACTER.

Animal Terebella, with two calcareous hemisphærical valves cut off before, and two lanceolate ones. Shell roundish, flexuous, and capable of penetrating into wood.

SPECIFIC CHARACTER

AND

SYNONYMS.

Shell very thin, cylindrical and smooth.

TEREDO NAVALIS: testa tenuissima cylindrica lævi. *Gmel. p.* 3747. 334. *sp.* 1.

TEREDO, *Linn. Syst. Nat.* 12. 2. *p.* 1267. *n.* 1.

A 2

PLATE CXLV.

Dentalium testa membranacea cylindracea, ligno inserta. *Linn. Fn. Suec.* 1. *p.* 380. *No.* 1329.

Serpula testa cylindracea flexuosa, lignum perforans. Teredo. *Da Costa. Brit. Conch. p.* 21. *sp.* 11.

Sellius Hist. Nat. Tered. Baster, Phil. trans. 61.

TEREDO NAVALIS. SHIP-WORM. *Penn. Brit. Zool.* 4. *No.* 160.

This destructive creature is supposed to have been originally a native of the East-Indies, and from thence introduced into the European seas: at present it may be considered with propriety as a naturalized British species; and it is a fortunate circumstance that it does not thrive so well with us as in warmer climates.

The animal, a soft and almost shapeless gelatinous body, is furnished with a calcareous process, or augur, at the head, with which it bores with the utmost facility into the stoutest oaken plank, as it lies in the water; and where a number of them attack the same piece of wood, will in a few days entirely destroy it: hence the ravages of these animals in the bottoms of ships are fraught with the greatest danger; and notwithstanding all the precaution of sheathing the bottoms of ships with copper, they insinuate themselves through the smallest cavities, and lodge themselves securely in the timbers. Where the work of the animal first commences, the shell is obtusely rounded and closed, and as it proceeds it continues to lengthen its shell till, as Gmelin says, it becomes from four to six inches in length;—we have seen one of them whose progress through the solid plank had not been interrupted, that had grown nearly to the length of eighteen inches. It is said that sheets of paper dipped in tar, and applied to the ship's bottom, will prove a more effectual preservative of the timber than the usual sheathing of copper, and an extensive manufactory has been of late established for the preparation of this article: how far it may prove ultimately successful we cannot presume to imagine, but perhaps both the paper and the copper might be employed together with greater advantage than either of those articles separately.

For a more complete history of the Teredo than we might have otherwise possessed, we are indebted to a remarkable circumstance that occurred about sixty years since: the piles on the coast of Holland were found to be injured to a very alarming degree, by the ravages of this creature; and beside several other ingenious tracts upon their history and the calamity they had occasioned, *Sellius* published an account of it, under the title of Historia Naturalis Teredines, seu, Xylophagi Marini, in 1733; in this book the anatomy of the animal is illustrated with Plates, and upon the whole his observations deserve the attention of the curious reader. Another account was also written by *Baster*, and published in the Transactions of the Royal Society of London, in vol. 61, as quoted above.

In our specimens, the apertures, or mouths of the shells, are very perfect, and exhibit the same appearance as *Kæmmer* and *Gmelin* seem to think peculiar to the species *Utriculus*; namely, an oval aperture divided by a partition in the middle. The shell is extremely delicate, or thin, and very brittle.

PLATE CXLVI.

PATELLA INTORTA.

INCLINING PATELLA, OR LIMPET.

GENERIC CHARACTER.

Animal Limax. Shell univalve subconic and without spire.

SPECIFIC CHARACTER

AND

SYNONYMS.

Shell entire, ovate, furrowed: ribs slightly imbricated, vertex somewhat reflexed and obtuse.

PATELLA INTORTA: testa integris ovata, sulcata: costis sub-imbricatis, vertice sub-reflexo obtuso.

PATELLA INTORTA, *inclining:* with an elevated shell, slightly striated; the vertex bending, but not hooked. *Penn. Brit. Zool.*

This shell is described by Pennant, who acquaints us it "inhabits Anglesea, found on the shores." It is a very rare shell, but has been taken also on the western coast, and communicated by J. Laskey, Esq. of Crediton, Devonshire.

The figure of Patella intorta, in the British Zoology, is certainly very indifferent: but having examined the shell Mr. Pennant de-

A 4

scribes, we have no hesitation in saying that it is not the Patella mammillaris of Gmelin, as some conchologists imagine. Specimens of the latter we are likewise in possession of, but they are not certainly known to be natives of this country.—In *Lister Conch.* *t.* 537. *fig.* 17; and in *Martini. Conch.* 1. *t.* 7. *f.* 58, 59. P. mammillaris is very accurately figured; and a slight comparison of either of them, with the shell figured in the annexed plate, will prove very clearly that they cannot be of the same species.

147

PLATE CXLVII.

PATELLA LACUSTRIS.

LAKE LIMPET.

GENERIC CHARACTER.

Animal Limax. Shell univalve, subconic, without spire.

SPECIFIC CHARACTER

AND

SYNONYMS.

Shell very entire, oval, membranaceous: crown pointed and reflected.

PATELLA LACUSTRIS: testa integerrima ovali membranacea: vertice mucronato reflexo. *Fn. suec.* 2200.—*Gmel. Syst. Nat. T.* 1. *p.* 6.

Patella fluviatilis, fusca, vertice mucronato, incurvo, inflexoque. *Gualt. Ind. Conch. tab.* 4. *fig. B.*

Patella fluviatilis, exigua, fubflava, vertice mucronato, inflexoque. *List. Hist. Conch. tab.* 141. *fig.* 39. *Morton Northamp. p.* 417.

PATELLA LACUSTRIS, *Penn. Brit. Zool.* 4. *No.* 149.

PATELLA integra, exigua, fusca, fragilis, vertice inflexo. *Da Costa. Brit. Conch.* 1. *tab.* 2. *fig.* 8. 8.

This is a thin and brittle shell, of a pale brown, or whitish colour, that is found on aquatic plants, in most ponds and rivers in

Europe : in England it is very common in some places. The animal, as Gmelin describes it, has two truncated and concealed tentacula, each of which is furnished with an eye at the inner angle.

Dr. Lister informs us, that they couple in September, and fix their spawn plentifully on stones and other bodies in the water : this spawn consists of little gelatinous globules, in each of which it is said many small shells may be distinguished.—The shell is shewn of the natural size in the annexed plate.

PLATE CXLVIII.

LEPAS TINTINNABULUM.

BELL ACORN SHELL.

GENERIC CHARACTER.

Animal Triton. Shell of many valves, affixed by a stem or broad base.

SPECIFIC CHARACTER
AND
SYNONYMS.

Shell conic, obtuse, rugged and fixed.

LEPAS TINTINNABULUM: testa, conica, obtusa, rugosa fixa. *Mus. Lud. Ulr.* 466. *n.* 3.

Balanus major angustus purpurascens, capitis apertura valde patente. *List. H. Conch. tab.* 433. *fig.* 285.

Balanus major. The conic centre shell. *Grew. Mus. p.* 148.

Balanus maximus ore patulo. *Mus. Petiv. p.* 82. *No.* 803.

B. tintinnabuliformis et B. calyciformis orientalis. *Phil. Trans.* 1758. *p.* 11. *tab.* 34. *fig.* 8, 9.

B. ore hiante magnus. *Borlase. Corn. p.* 27.

Gland de Mer clochette, *D'Avila Cab. p.* 404. *No.* 922.

BALANUS TINTINNABULUM BELL. B. major purpurascens, conicus, angustus tintinnabuliformis, apertura valde patente. *Da Costa Brit. Conch. p.* 250. *sp.* 70.

BALANUS TINTINNABULUM BELL. *Penn. Brit. Zool. T.* 4. *No.* 8.

PLATE CXLVIII.

This kind of Balanus is found affixed in large clusters to the bottoms of ships in our seas, but the general opinion is that it originates in warmer climates, and should not be considered an indigenous British species. A supposed variety of it of a dirty whitish colour, is said to be found in the North seas, by Chemnitz.

Balanus Tintinnabulum is admitted among the testaceous productions of our seas by Borlase, Pennant and Da Costa, and this we must confess is the best apology we have to offer for inserting it in the present Work.

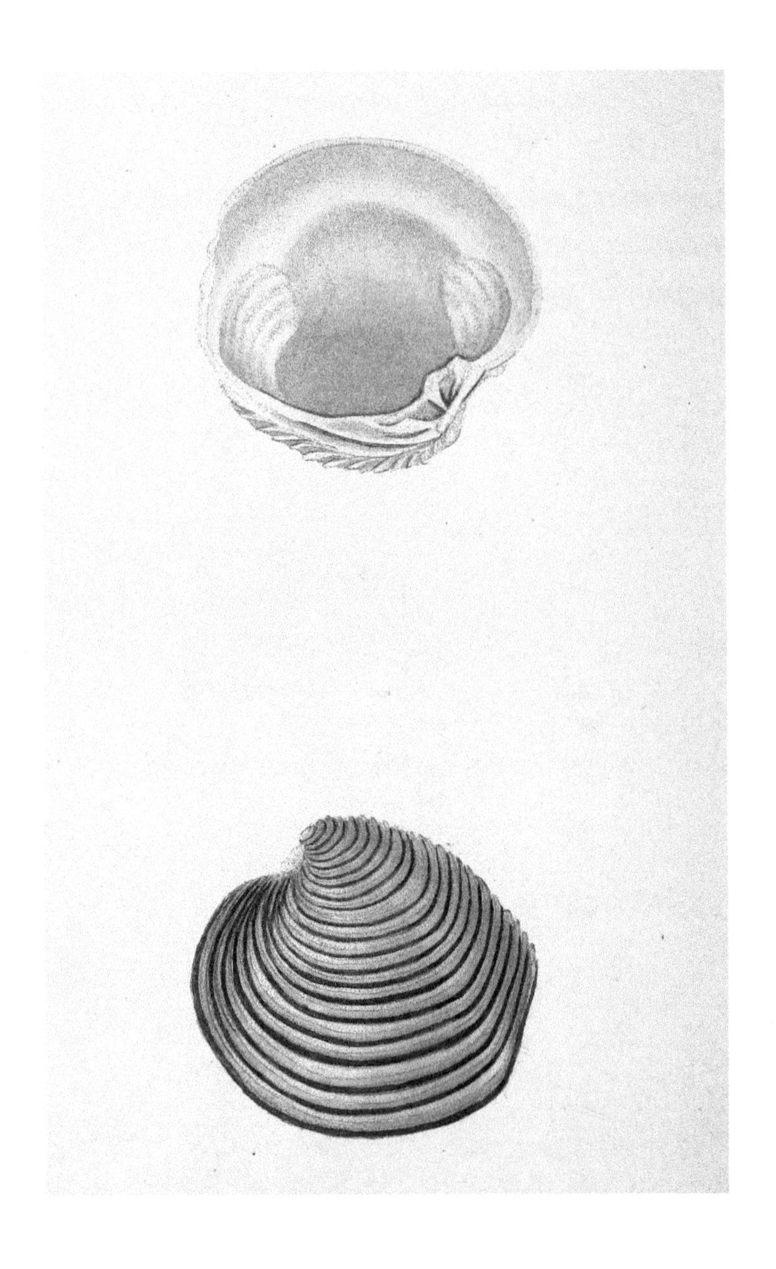

PLATE CXLIX.

VENUS LACTEA.

MILKY VENUS SHELL.

GENERIC CHARACTER.

Bivalve. Hinge furnished with three teeth; two near each other, the third divergent from the beaks.

SPECIFIC CHARACTER.

Shell lentiform, somewhat compressed, with thick, elevated, obtuse concentric striæ, and slightly truncated anteriorly.

VENUS LACTEA: testa lentiformi sub-compressa: striis concentricis crassis elevatis obtusis, antrorsum subtruncata.

This appears to be a new and undescribed British species of Venus, approaching, in some particulars, to others of the same genus found on our coast, although differing in having the concentric striæ or ridges large, elevated, and obtusely rounded. The striæ, for example, in V. borealis, and V. cancellata rise in a thin membranaceous ridge to an acute edge; and the former of these seems at first sight to bear a strong analogy to our shell: Venus lactea is also a much thicker and heavier shell than any other resembling it, with which we are acquainted. V. Exoleta has thick, but minute striæ.

Our present species, we are informed, is found on the western coast.

PLATE CL.

PATELLA OBLONGA.

OBLONG FRESH WATER PATELLA, OR LIMPET.

GENERIC CHARACTER.

Animal Limax. Shell univalve, sub-conic, and without spire.

SPECIFIC CHARACTER

AND

SYNONYMS.

Shell very entire, oblong, compressed, membranaceous: vertex pointed and reflected obliquely, or to one side.

PATELLA OBLONGA: testa ingerrima oblonga compressa membranacea, vertice mucronato reflexo oblique. *Lightfoot. Phil. Trans. V.* 76. *p.* 167.

This species was first described in the transactions of the Royal Society of London, by the Rev. Mr. Lightfoot, chaplain to the late Duchess of Portland. He says it was found adhering to the leaves of the *Iris Pseudacorus*, in waters near Beaconsfield in Buckinghamshire, by Mr. Agneu, the Duchess of Portland's Gardener. It has been since found on plants in the river Stour, by the Rev. Thomas Rackett.

PLATE CL.

It is evidently distinct from the *Patella lacustris* of Linnæus, in being of an oblong instead of ovate form ; and in having the pointed vertex bending obliquely or to one side, instead of being centrical and reflected back. The colour is variable, in some it is greenish, and in others of a pale brown. It is represented on our plate both of the natural size and magnified.

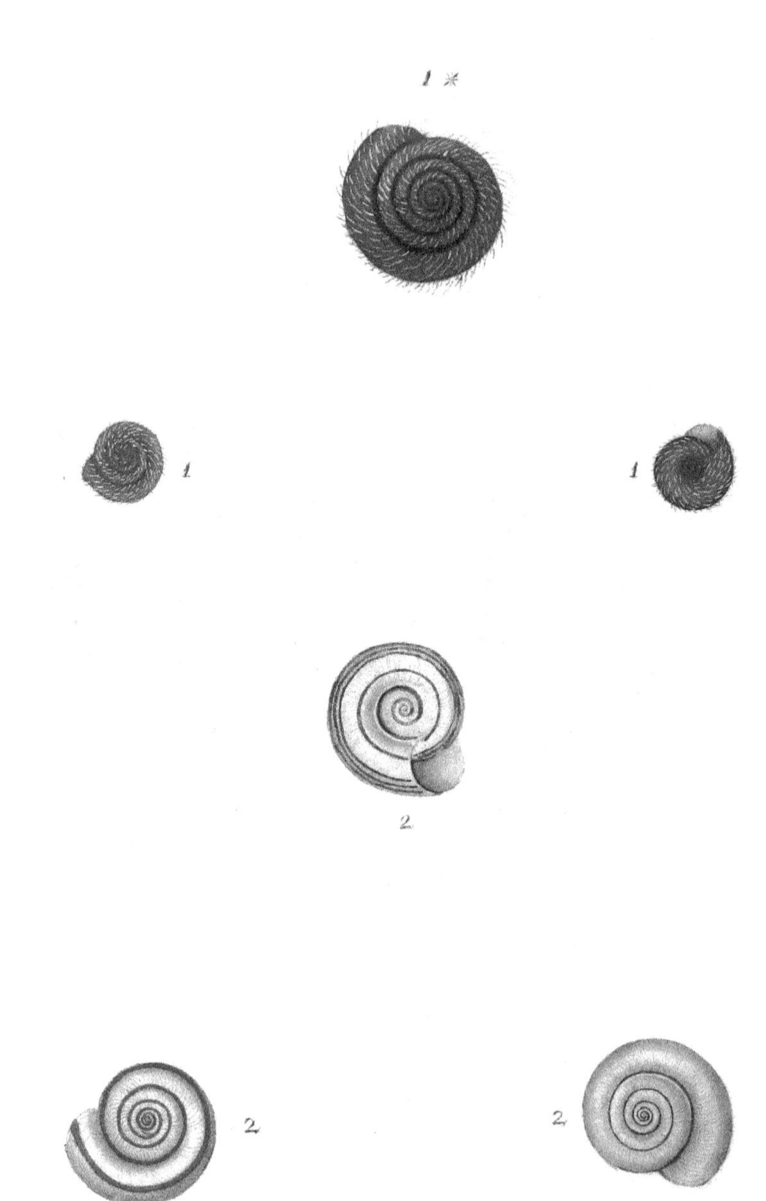
1 *
1
1
2
2
2

PLATE CLI.

FIG. I.

HELIX HISPIDA.

BRISTLY SNAIL.

GENERIC CHARACTER.

Animal Limax. Shell univalve, spiral, diaphanous, fragile. Aperture contracted semilunar, or roundish.

SPECIFIC CHARACTER
AND
SYNONYMS.

Shell umbilicated, convex, hairy, diaphanous; whorls five: aperture roundish-lunated.

HELIX HISPIDA: testa umbilicata convexa hispida diaphana: anfractibus quinis, apertura subrotundo-lunata. *Linn. Fn. Suec.* 2182.—*Gmel. Syst. Nat.* 3625.

Helix sub-globosa umbilicata, cornea, diaphana, hispida. *Hispida,* Helix. *Da Costa. Brit. Conch. p.* 58.

This shell is not unfrequently found at the bottoms of trees among the moss, in woods and wet shady places. It is glossy, very thin, brittle, and of a brown horny colour. When the animal is alive in the shell it is of a dark red colour approaching to black, and is very elegantly set all over with minute, short, white bristles, or hairs, which easily rub or fall off when the animal dies.

In the plate the upper and underside are represented, together with a magnified figure, which is distinguished by a star, and is intended to shew the hispid appearance of the shell while the animal is alive.

FIG. II.

HELIX ERICETORUM.

HEATH SNAIL.

SPECIFIC CHARACTER
AND
SYNONYMS.

Shell umbilicated, depressed and yellowish, with one or more fuscous bands.

HELIX ERICETORUM: testa umbilicata depressa lutescente: fascia una vel pluribus fuscis. *Müll. Hist. Verm.* 2. *p.* 33. *n.* 236.—*Gmel. Linn. Syst. Nat. p.* 3632.

Cochlea cinerea albidave, faciata, ericetorum. *List. H. An. Angl. p.* 126. *tit.* 13. *tab.* 2. *f.* 13.—Cochlea compressa, umbilicata fasciata campestris. *List. H. Conch. tab.* 78. *fig.* 78.

PLATE CLI.

HELIX ALBELLA, *Penn. Brit. Zool. Vol.* 4. *tab.* 85. *fig.* 122.
HELIX cinerea albidave, fasciata, ericetorum Erica. *Da Costa. Brit. Conch. p.* 53. *Sp.* 30.

This species of Helix, as its name implies, is found on heaths and sandy soils, and is very common both in this country and other parts of Europe. When full grown, this shell is three quarters of an inch in breadth, and one-third of its breadth in height: the spires flat: the outermost wreath very convex beneath, with a large and deep central umbilicus; and circular mouth or aperture.

The young shells are quite plain, and of a horny colour, or whitish and semitransparent. When full grown they are opake, dull, white or yellowish, and usually fasciated with one or more brown circular bands, according to the involutions of the wreaths. The order, size, and number of these brown bands, as Da Costa says, vary extremely, though commonly there is one band in the middle or near the bottom of each wreath, and often other fainter and narrower bands accompany it. Gmelin speaks of five distinct varieties, which differ in size, in colour, and number of the bands. Sometimes they are quite white, or marked with a single spiral band; and sometimes these bands amount to eight or nine on each shell.

Dead shells of this kind are found in vast numbers intermixed with the sand on heaths, and are always observed in great plenty with the others.

PLATE CLII.

PINNA LÆVIS.

SMOOTH PINNA, OR HAM SHELL.

GENERIC CHARACTER.

Animal Limax. Shell sub-bivalve, fragile, erect, gaping at one end, and furnished with a byssus or beard: hinge toothless, and uniting the valves into one.

SPECIFIC CHARACTER.

Shell nearly triangular, horn-colour, smooth: valves rugose on the posterior part.

PINNA LÆVIS: testa sub-triangulari cornea lævi: valvis posterius rugosis.

This species of Pinna which differs from any that has been before described as a British shell; and if we are not mistaken, from either of the Linnæan or Gmelinian species of the genus also; was received by A. M'Leay, Esq. among other curious shells that were dredged up on the coast of Shetland.

The difference between this and the other analogous kinds, seems to consist in its being of a more triangular form, and in not having the least trace of spines or murication: from the beak descend some very obsolete longitudinal striæ, but the surface is in general perfectly smooth and glossy, notwithstanding the specimen before us has at first sight a rugged aspect; the shell having been greatly bruised or mutilated in its growth, and afterwards uncouthly repaired by the animal inhabitant.

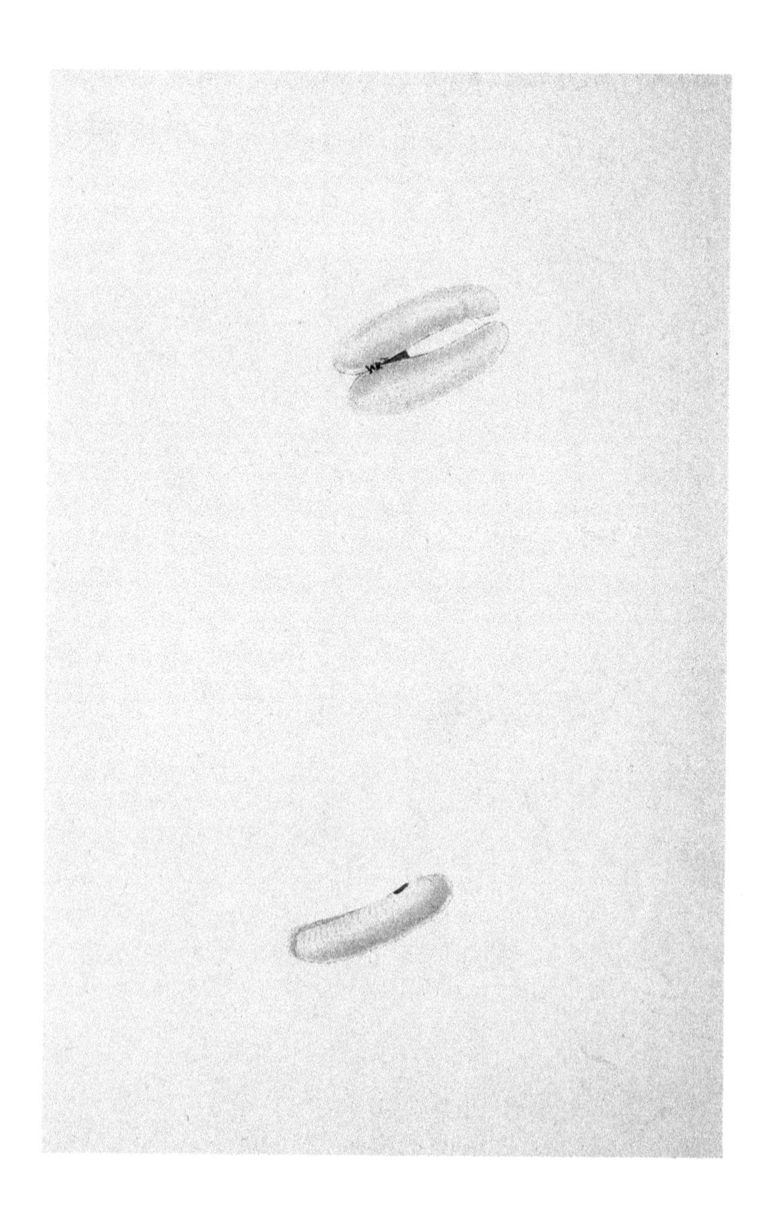

PLATE CLIII.

SOLEN PELLUCIDUS.

PELLUCID RAZOR SHELL.

GENERIC CHARACTER.

Bivalve, with equal valves, oblong, open at both ends; at the hinge a subulated tooth turned back, often double; not inserted in the opposite shell. Animal an ascidia.

SPECIFIC CHARACTER AND SYNONYMS.

Shell suboval, somewhat arcuated, fragile, pellucid: hinge with an acute bidentated tooth on one side.

Solen Pellucidus: testa subovali subarcuata fragile pellucida cardine altero acute bidentato.

Solen Pellucidus, *Penn. Brit. Zool. T. 4. p. 1. p. 84. sp. 23.*

A very rare species, and described only by Mr. Pennant, who says it inhabits Red Wharf, Anglesea.

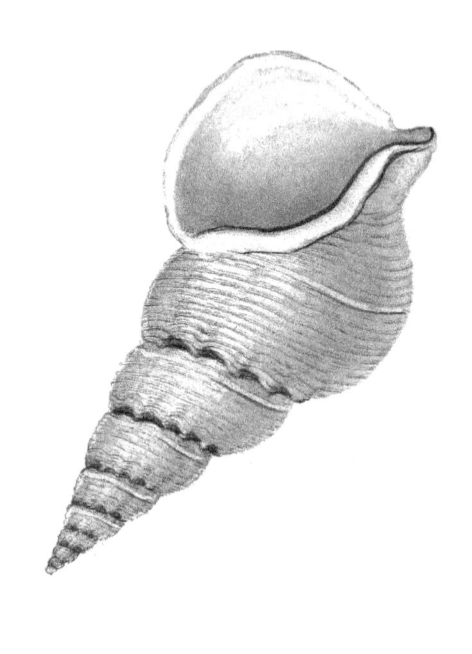

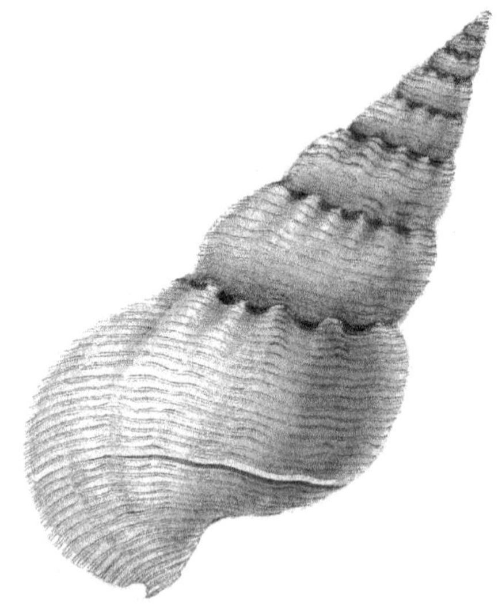

PLATE CLIV.

BUCCINUM GLACIALE.

NORTHERN WHELK.

GENERIC CHARACTER.

Animal Limax. Shell univalve, spiral, gibbous. Aperture ovate, ending in a short canal or gutter, inclining to the right: tail or beak retuse: inner lip expanded.

SPECIFIC CHARACTER
AND
SYNONYMS.

Shell smooth, somewhat striated, ovate-oblong; lower whorl slightly carinated.

BUCCINUM GLACIALE: testa lævi substriata ovato-oblonga: anfractu infimo subcarinato. *Linn. Faun. Suec.* 2162.—*Gmel. Syst. Nat.* 4. 3491. *sp.* 92. *Chemn. Conch.* 10 *t.* 152. 1446, 1447.

TRITONIUM GLACIALE, *O. Fabr. Faun. Grænl. No.* 397.

As a native of the North Seas, this shell was well known to Linnæus, by whom it was most accurately and minutely described in his

Fauna Suecica *, but we have no other authority for believing it to be an inhabitant of our own seas, than that of Mr. Agneu, gardener to the late dutchess of Portland, by whom it was discovered among the Orkney Islands, and, in consequence, admitted into the collection of British shells in the Portland Museum.

One of the most striking characters of this shell, is a single carinated ridge that surrounds the first or largest wreath of the shell, and does not afterwards appear on either of the rest. This it may be proper to notice, since the circumstance has been strictly mentioned both by Linnæus, Fabricius, (*Faun. Groen.*) and Chemitz, but it is not certainly a constant criterion of the species: there was a variety of this kind in the collection of the late Dr. Fordyce, at present in that of the Earl of Tankerville, in which the carinated ridge distinctly traverses the whole shell in a spiral course, from the first wreath nearly to the apex.—The latter was from Newfoundland.

* Buccinum glaciale; testa crassa magnitudine extimi articuli pollicees, pallida, secundum anfractus obsolete striata, acuminata superne conica. Anfractus infimus seu maximus subcarinatus est, sed hæc carina in reliquis superioribus anfractibus evanescit, cum sutura anfractuum evadat, quæ attenuata. Basis gibba emarginata. Apertura ovata. Labium exterius crassum patulum, striis incumbentibus. Linn. Fn. Suec.

1

1

1 ×

2

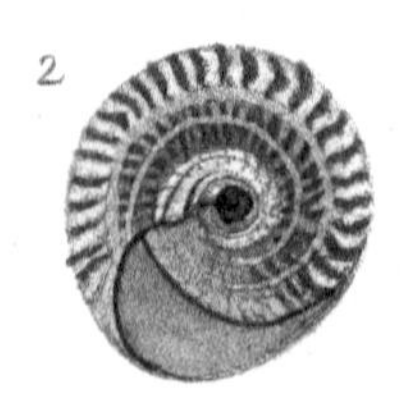

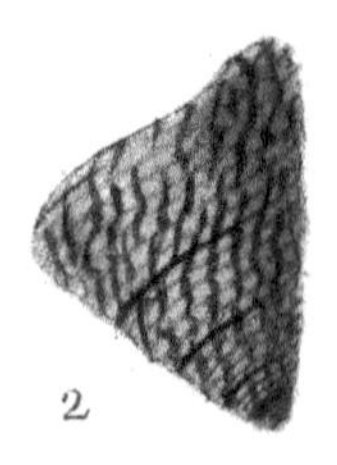

2

PLATE CLV.

FIG. I. I.

TROCHUS CONICUS.

CONIC TROCHUS, OR TOP SHELL.

GENERIC CHARACTER.

Animal a Slug. Shell conic: aperture nearly triangular.

SPECIFIC CHARACTER

AND

SYNONYMS.

Shell conic, smooth, whitish, obliquely lineated with brown, whorls flattish, and finely striated.

TROCHUS CONICUS: testa conica lævi, albida oblique fusco lineata: anfractibus planiusculis subtiliter striatis.

A small shell bearing some affinity with Trochus Conulus, from which it is notwithstanding perfectly distinct; as a British shell we believe it is altogether new; nor does it seem to be described by any foreign author. Four of these shells were picked up on the sea coast of Devonshire by J. Laskey, Esq. from whom we received the specimens represented in our Plate. It has been since communicated also from the Mediterranean sea.

D 2

PLATE CLV.

FIG. II. II.

TROCHUS CINEREUS.

ASHEN TOP SHELL.

SPECIFIC CHARACTER.

Shell pyramidal, umbilical, cinereous; marked with narrow blackish lines.

TROCHUS CINEREUS: pyramidalis umbilicatus, cinereus, lineis angustis nigrescentibus notatus. *Da Costa. Brit. Conch. p.* 42. *sp.* 23. *tab.* 3. *fig.* 9. 10.

Trochus pyram, parvus, ex viridi sive subcæruleo variegatus, insigniter umbilicatus. *List. H. Conch. tab.* 633. *fig.* 21.

This shell is described and figured by Da Costa from the specimen at present in our possession; the only inducement we have for inserting it, for though this writer observes that it is a common shell on several of our coasts, we must acknowledge it has never occurred to us as a British shell. Exotic specimens we have, but these are said to have been brought from the South Seas. Da Costa, we have a strong suspicion was mistaken concerning this shell; his reference to Lister is correct; the rest of his synonyms, namely, those of *Dale*, *Wallis*, &c. are erroneous, since those writers meant a very different shell. The following is the minute description Da Costa gives of this kind:

" The shell is thick and strong, of the size of a cherry; shape obtusely pyramidal, or not quite tapering to a point.

PLATE CLV.

" The base is very concave, with some circular furrows; the mouth roundish and capacious, *within* fine mother-of-pearl; the *outer lip* smooth and even ; the inner or *pillar lip* has two jags or slight teeth, and two furrows crossing it transversely ; from hence it widens, runs oblique, and forms a spacious cavity, at the bottom of which lies the *umbilicus*, deep, cylindric, and so hollow as to admit the head of a large pin. All this part is of a dark ash, greatly variegated with blackish lines, or streaks, which run lengthways and across ; but the beginning of the umbilicus is generally pearly, and of a fine light greenish colour.

" The body and turban have five bellied, or swelled wreaths, or whorls, separated by a very depressed line ; they are circularly striated, but faintly, and the colours are exactly the same as at the base."

156

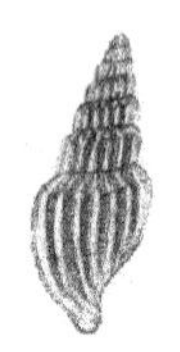

PLATE CLVI.

MUREX ANGULATUS.

ANGULATED MUREX.

GENERIC CHARACTER.

Animal Limax. Shell spiral, rough: aperture ending in a straight, and somewhat produced gutter or canaliculation.

SPECIFIC CHARACTER.

Shell oblong, whorls depressed, angulated, transversely striated, sulcated longitudinally; aperture toothless.

MUREX ANGULATUS: testa oblonga: anfractibus depressis angulatis transversim striatis longitudinaliter sulcatis, apertura edentula.

A neat shell of interesting figure, that has been found, though rarely, on the English coasts. The specimens figured in our Plate were discovered on the sands at Brighton by Mr. Munn, who kindly communicated them to us. We have received it since from the coast of Weymouth.

This is doubtless an undescribed species.

D 4

157

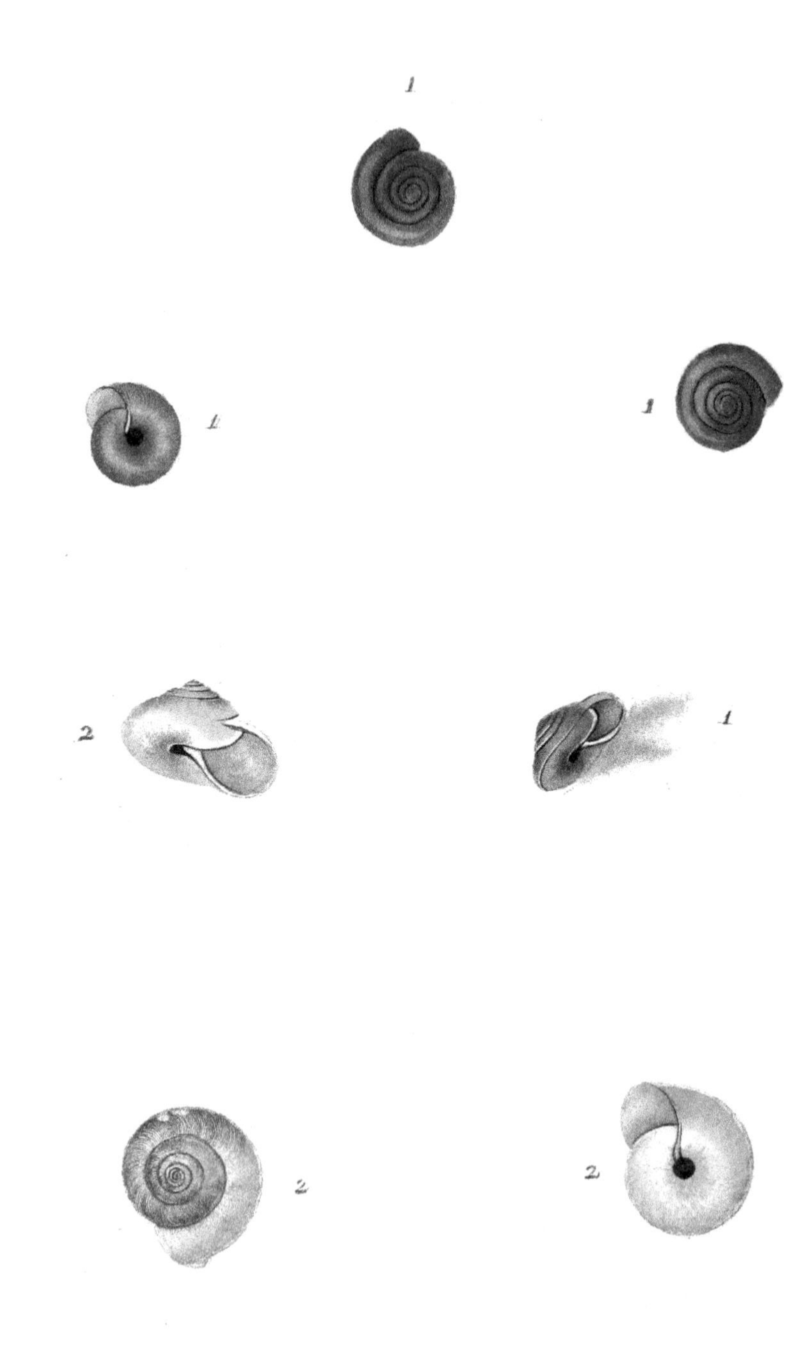

PLATE CLVII.

FIG. I. I.

HELIX RUFESCENS.

REDDISH SNAIL.

GENERIC CHARACTER.

Animal Limax. Shell univalve, spiral, diaphanous, fragile. Aperture semilunar, or roundish.

SPECIFIC CHARACTER
AND
SYNONYMS.

Shell umbilicated, and streaked with pale reddish.

HELIX RUFESCENS: testa umbilicata, et striata dilute rufescens.

Cochlea umbilicata, et striata dilute rufescens. RUFESCENS *Da Costa. Brit. conch. p.* 80. *sp.* 43.

Cochlea dilute rufescens, aut subalbida, sinu ad umbilicum exiguo, circinata. *List. H. An. Angl. p.* 125. *tit.* 12. *tab.* 2. *fig.* 12.

Cochlea terrestris depressa & umbilicata mellei coloris, labio candido repando, sinu ad umbilicum exiguo circinato. *Gualt.* 1. *Conch. tab.* 3. *fig. N*

PLATE CLVII.

This kind may be readily distinguished by a slight carene, or ridge that surrounds the first or largest wreath of the shell. The colour, as the name implies, is reddish when the animal is alive; when dead, whitish and discoloured: the carene is usually of a lighter colour than the rest of the shell.

Da Costa speaks of it as being not very common; observing at the same time that he had received it from Cornwall and Hampshire. It is pretty frequent, he adds, about Leeswood in Flintshire, between the bark and wood of trees thrown down, and decayed, especially alders. Dr. Lister found it in plenty about Tadcaster, in the woods and hedges of marshy and shady meadows, and in like places throughout Craven, in Yorkshire: he observes there is a variety, (if not a different species,) in Kent, somewhat larger, lighter coloured, and with a smaller umbilicus. Mr. Morton found it at Morsley, and the other, Northamptonshire woods. To this we should add, that from our own observation, the species appears to be more frequent in many parts of the country than our author imagined. It delights chiefly in marshy places. Occasionally we have found it on aquatic plants in Battersea meadows.

FIG. II.

HELIX PALLIDA.

PALE SNAIL.

SPECIFIC CHARACTER.

Shell inflated, slightly umbilicated, fragile, pale: whorls six, convex; aperture semilunar.

PLATE CLVII.

HELIX PALLIDA: testa inflata subumbilicata fragili pallida: anfractibus sex convexis, apertura semilunari.

Rather a local species, found in some parts of Kent: we have also seen it on the great roman wall of Caerwent, Monmouthshire.

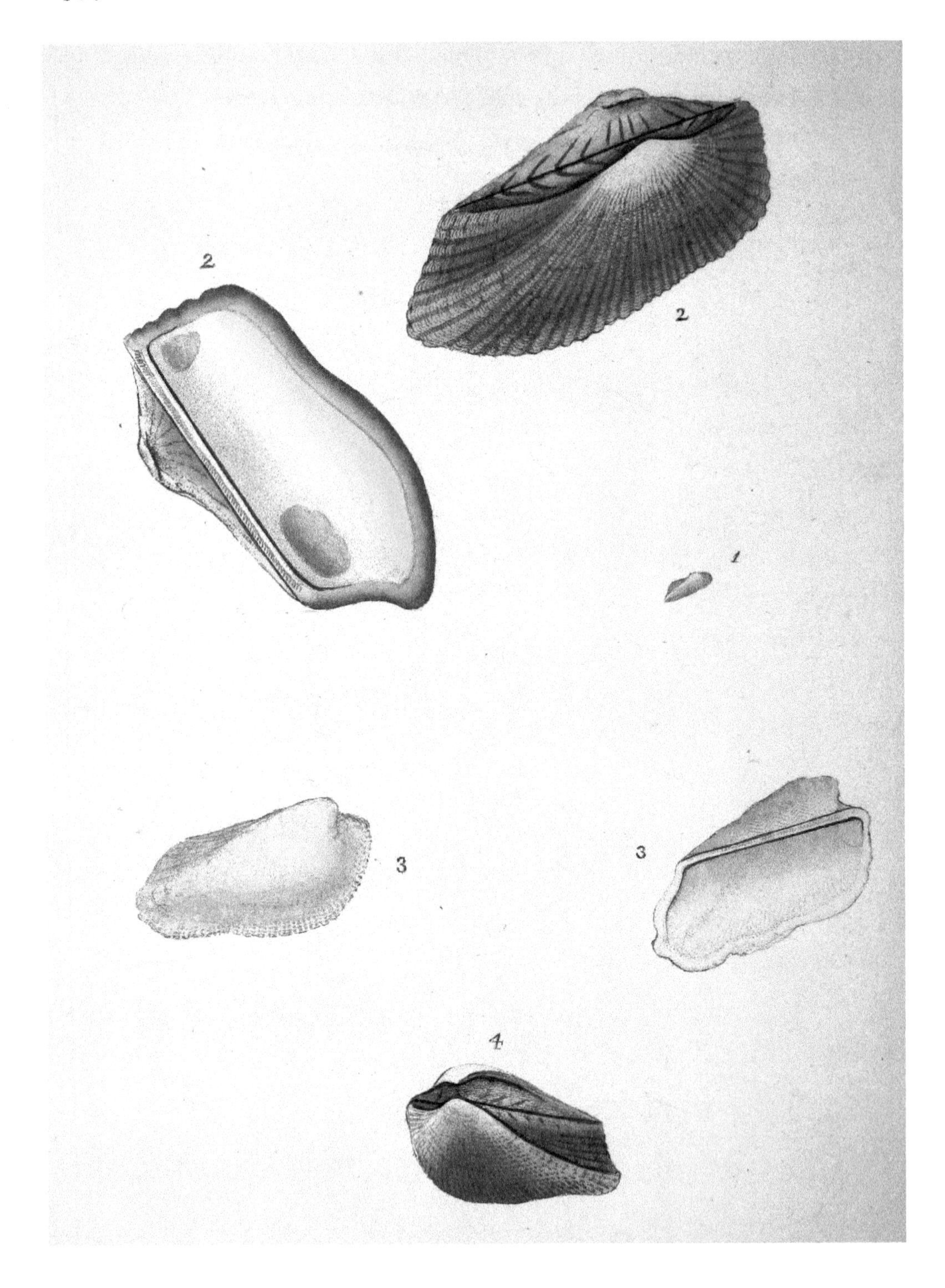

2
2
1
3
3
4

PLATE CLVIII.

FIG. I. II.

ARCA NOAE.

NOAH'S ARK SHELL.

GENERIC CHARACTER.

Shell bivalve, valves equal: teeth at the hinge numerous, acute, and inserted between each other.

SPECIFIC CHARACTER.

Shell oblong, striated, at the apex emarginated: beaks very remote: margin gaping.

ARCA NOAE: testa oblonga striata apice emarginata, natibus remotissimis, margine hiante. *Linn. Gmel. Syst. Nat. T.* 1. *p.* 6. *p.* 3306. *sp.* 2.

Musculus Matthioli. *List. t.* 368.

Concha rhomboides, *Rondel.* 99. 2. *p.* 27.
Bonanu. recr. 2. *f.* 32.

Miss Pocock, whose liberal communication of British shells collected by herself, we have had such frequent occasion to mention in the progress of this publication, has obliged us with a small specimen of the Arca Noae found on the shores of Cornwall; the smallest shell figured in the upper part of our Plate.

PLATE CLVIII.

It has been presumed before that this shell was a native of our coasts: that Borlase had met with it; and that the Arca tortuosa * of Pennant, (which he says inhabits Cornwall, and has been found near Weymouth,) was no other than the Arca Noae of Linnæus. But this still remained a matter of much uncertainty among Conchologists, and with ourselves, till we received the specimen from the lady above-mentioned, which proves beyond dispute that it is a British shell, and perfectly similar to those of the same species found in the Mediterranean sea. This being a young shell, is not of course covered with the rude brown epidermis, as in the old or full grown Mediterranean shell, figured with it, in order to elucidate the species with more precision. Since the publication of the Plate we have also had the satisfaction of receiving another specimen of the shell nearly thrice the size of the small one figured, in a parcel of shells collected on Slapton sands, Devonshire, last summer.

At the same time that Miss Pocock discovered this small specimen of Arca Noae, several worn valves of an Arca, confessedly of a different kind, occurred likewise. Both the internal and external view of these are represented in the lower part of the plate, fig. 3, together with that of the perfect specimen of a foreign shell, fig. 4, which may prove hereafter to be of the same species. The mutilated valves we have little hesitation in believing it to be precisely those of the shell figured by Lister, *t.* 367. *n.* 207. *Balanus Bellonij tenuiter striata*; though from their imperfect condition it might be improper to offer any positive opinion concerning them. At a future period we may be enabled, by receiving better specimens, to ascertain this point, and as it may then appear, our conjectures were not unfounded.

* No. 57, Penn. Brit. Zool.

PLATE CLVIII.

We should further add, that this species, though observed by Lister, has been overlooked by Linnæus; and that from a MS. note in one of the copies of Lister's work, in the library of Sir Joseph Banks, we find the late Dr. Solander intended to have named it specifically *fusca*, had he lived to publish his new arrangement of Conchology.

*

PLATE CLIX.

TURBO RETICULATUS.

RETICULATED TURBO.

GENERIC CHARACTER.

Animal Limax. Univalve, spiral, or of a taper form. Aperture somewhat compressed, orbicular, entire.

SPECIFIC CHARACTER.

Shell tapering, reticulated with granules, testaceous, whorl reversed; aperture straitened.

TURBO RETICULATUS; testa turrita reticulata granulata testacea, anfractibus coarctata.

This is a remarkably neat, or rather elegant shell, and equally distinguished for its rarity. It was found in the sands on the coast of Cornwall, by Miss Pocock, to whose polite attention we are exclusively indebted for the specimen now figured.

That this small species of Turbo is undescribed either as a British or a foreign shell, we have little reason to dispute, unless the following, described by Mr. Walker, should prove to be the same: "Turbo, Turritus perversus novem anfractibus punctatis apertura coarctata; the reversed taper Turbo of nine dotted whorls and straitened aperture.

A shell found at Sandwich."—The description does not strictly correspond with our shell, the figure is yet more remote. Some degree of ambiguity arises from this particular circumstance; in our shell the wreaths are uniformly lineated spirally, with three prominent rows of tubercles, or more correctly speaking, granulations, except on the first wreath, where they are more numerous, and the intermediate series on every wreath, consists of smaller granulations than those on either side of it. The term *punctatis*, on the contrary, which Mr. Walker has adopted, must rather imply a dot depressed: in the engraving also, by which his description is elucidated, the dots appear to be disposed in three distinct series upon each wreath, as the granulations are in the shell before us, but each dot is apparently depressed, and situated in the center of a quadrangular compartment: at the same time also it must be remarked, that the intermediate series of these dots on every wreath, are of an equal magnitude with the others. The aperture, whatever might be the shell designed, is miserably expressed, as are indeed the figures both of the natural size, and magnified. We suspect upon the whole, they are intended for our shell, and were it not for the objections stated, should insert a reference to his figure, plate 3. No. 48. as a synonym.

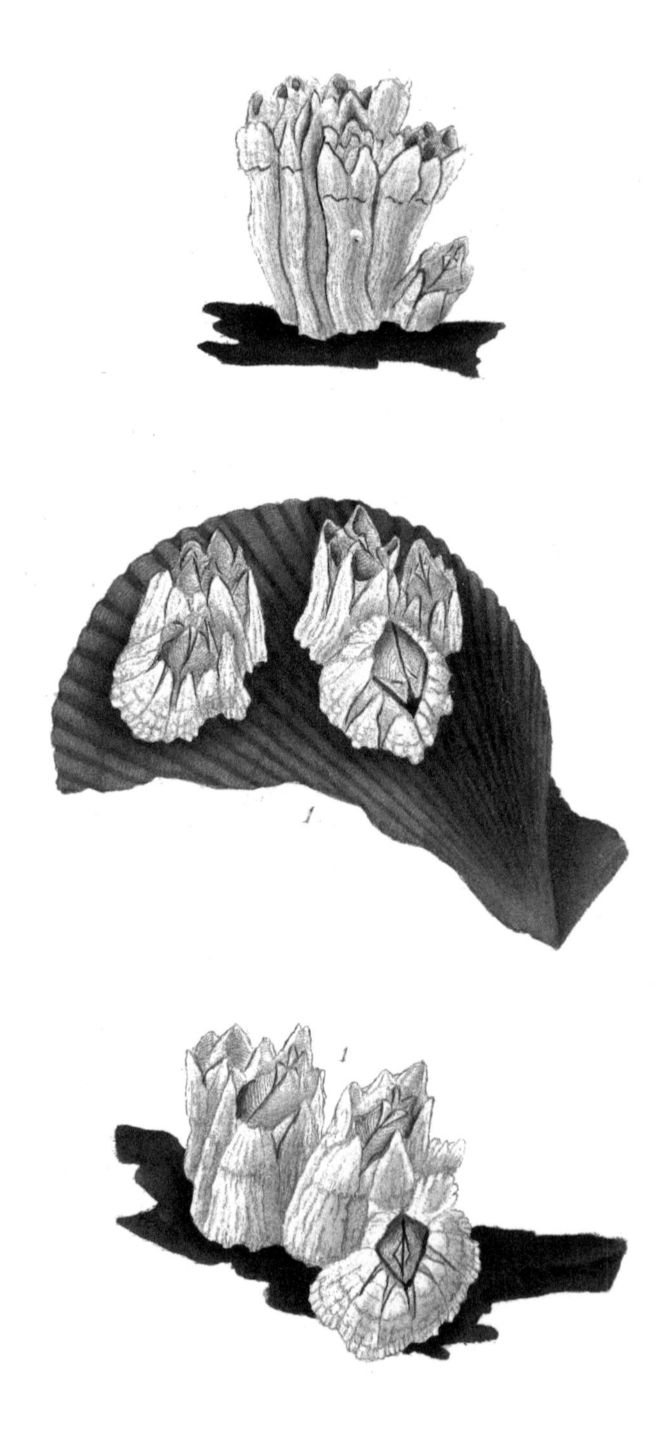
1
1

PLATE CLX.

LEPAS BOREALIS.

NORTHERN ACORN SHELL.

GENERIC CHARACTER.

Animal Triton. Shell affixed at the base; multivalve; the valves unequal.

SPECIFIC CHARACTER.

Shell erect, subconic, aperture quadrangular, operculum or lid acute, and striated transversely.

Lepas Borealis; testa erecta subconica, apertura quadrangulari operculis acutis transversim striatis.

A few small clusters, with some single specimens of this curious species of Lepas, were discovered about three years since, attached to the bottom of the Warspight ship of war, when taken into dock to be repaired, after lying in the harbour of Portsmouth for a great length of time. Mr. James Hay, of Portsmouth, has since found two or three shells of the same kind, by dredging in Portsmouth harbour; so that though probably not indigenous, it has now become a naturalised species.

This was first communicated by J. Laskey, Esq. We have since received it from Newfoundland, affixed to the valve of

E 2

a northern *ostrea*; and learning that it is undoubtedly a native of the north seas, we venture to assign it the specific name of *borealis.*—The clusters of these shells, when pressed together, (which rarely happens) take an elongated form, as is expressed in the upper figure.—We are certainly to consider this as a rare species.

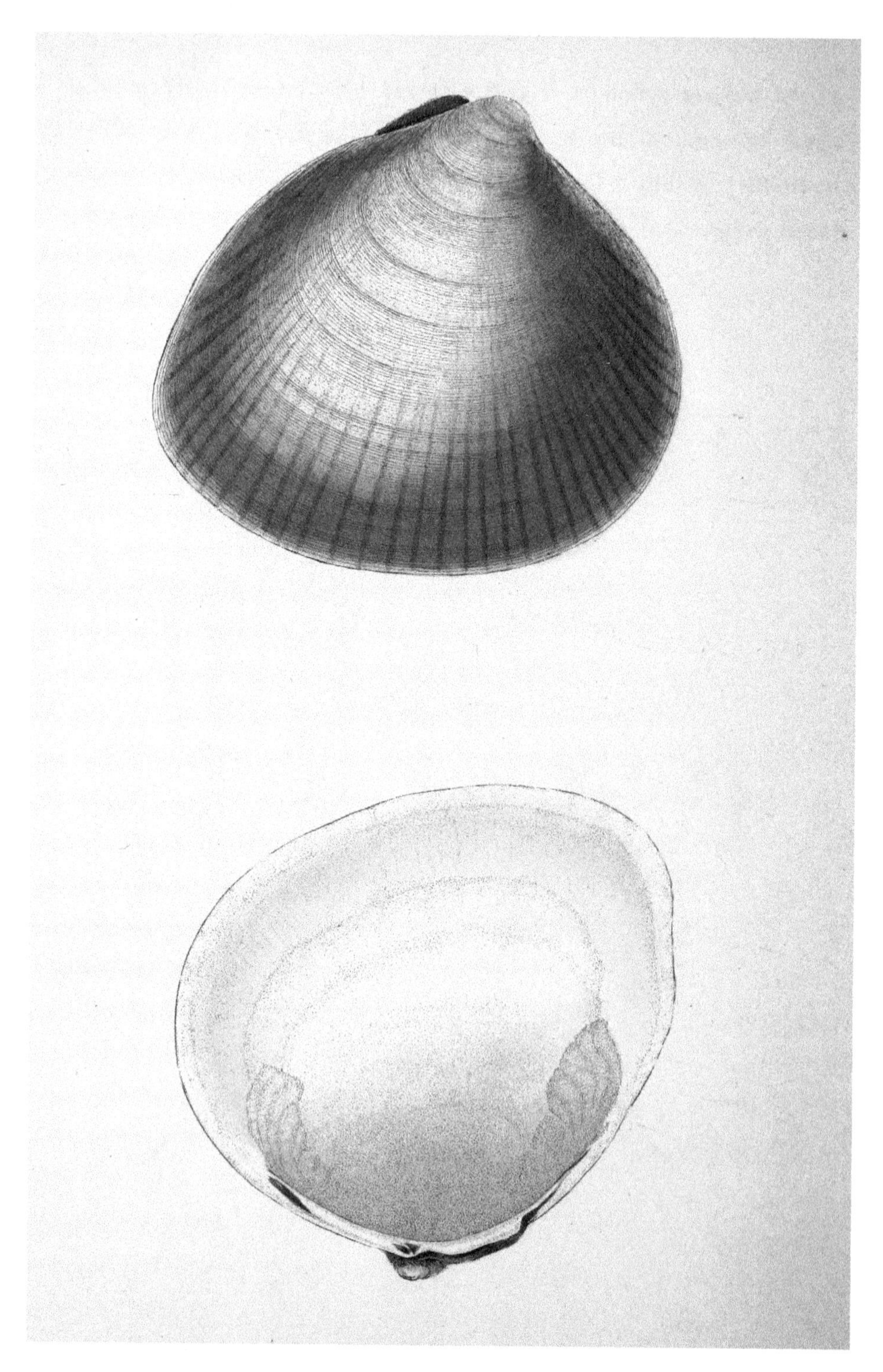

PLATE CLXI.

MACTRA RADIATA.

RADIATED MACTRA.

GENERIC CHARACTER.

Animal a Tethys. Shell bivalve, sides unequal. Middle tooth complicated, with a little groove on each side; lateral tooth remote.

SPECIFIC CHARACTER
AND
SYNONYMS.

Shell thin, fragile, somewhat triangular, compressed, whitish, and finely striated with testaceous rays.

MACTRA RADIATA: testa tenui fragili subtriangulari compressa albida subtilissime striata testaceo radiata.

Several valves of this shell were found upon Langston beach, near Portsmouth, after a severe storm that happened in the year 1800, by J. Laskey, Esq. of Crediton; from whom we received the specimens figured in the annexed plate.

This is a thin, brittle shell, of a large size; colour sordid white, tinged with reddish, and faintly marked with rays of a testaceous colour, beneath a brown filmy epidermis.

E 3

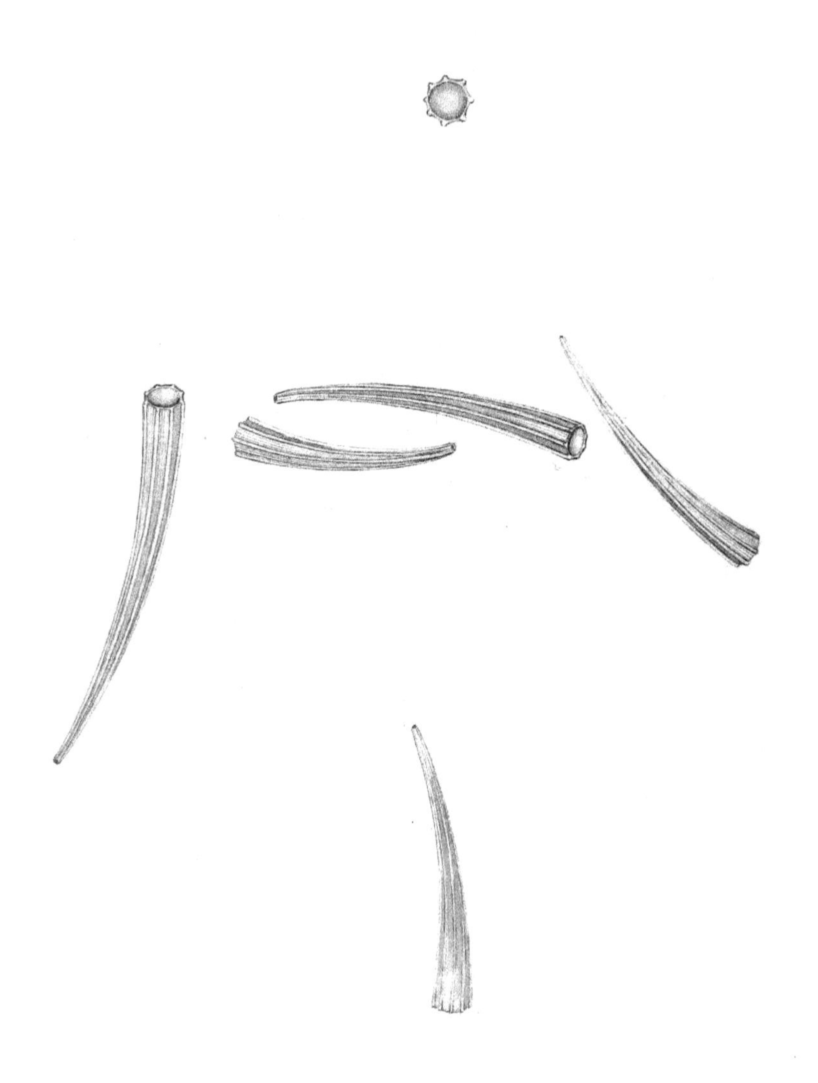

PLATE CLXII.

DENTALIUM OCTANGULATUM.

EIGHT-RIDGED TOOTH-SHELL.

GENERIC CHARACTER.

Animal a Terebella. Shell univalve, tubular, straight or slightly curved, with an undivided cavity open at both ends.

SPECIFIC CHARACTER AND SYNONYMS.

Shell white, somewhat curved, with eight ribs or angles, and three intermediate striæ.

DENTALIUM OCTANGULATUM: testa alba subarcuata octangulata: interstitius tri-striatis.

DENTALIUM STRIATULUM: *Gmel. Syst. Nat.* 3738. *sp.* 13?

For the discovery of this elegantly striated tooth-shell, as a native of the British coasts, we have once again to acknowledge our obligation to a lady, mentioned on other similar occasions in the progress of this work, Miss Pocock; several shells of this kind were found by her on the sandy coasts of Cornwall, near Lelant, in the year 1802.

PLATE CLXII.

It remains to express some little doubt, whether every circumstance will allow us to consider this as an hitherto undescribed species; as a British shell we can have no hesitation in saying it has not been mentioned by any author. We were rather inclined at first to think our shell could be no other than a variety of the Dentalium striatulum of Gmelin, which is described as a native of the Mediterranean and Sicilian seas. The synonyms given by Gmelin to that species, we found however to be less expressive of our shell than his description; Lister's shell, to which he refers, *t.* 547. *f.* 1. *b.* is much larger than our shell, as is likewise that of Martini, quoted with it; both are described to be of a fine green colour, with the tip only white, but it has uniformly eight distinct ribs or angles, as in our shell, which is a much more important characteristic of a species, than the mere difference of colour.—There is another shell, figured by *Martini, pl.* 1. *f.* 4. *B.* that seems to approach much nearer to our shell, and the colour is white, but as in D. elephantium, this has ten ribs instead of eight; it is the Dentalium aprinum of Gmelin.—As our shell, upon the whole, does not strictly accord with those species noticed, nor any others which we are acquainted with, a new name and character will tend at least to obviate confusion. Of Dentalium striatulum it may prove to be a variety, but that is doubtful; and there is scarcely any reason to dispute its being undescribed, unless it be of that species.

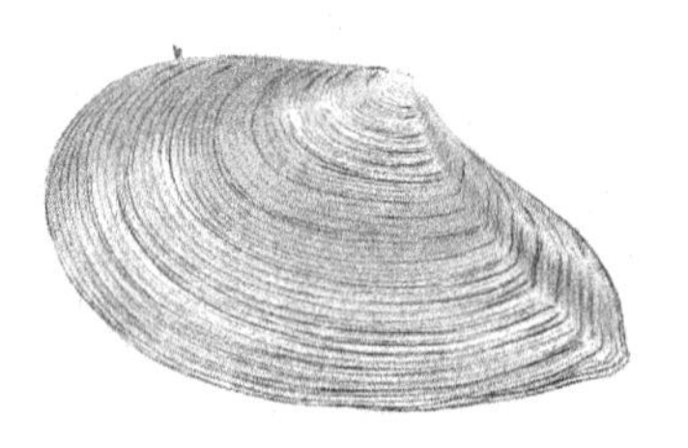

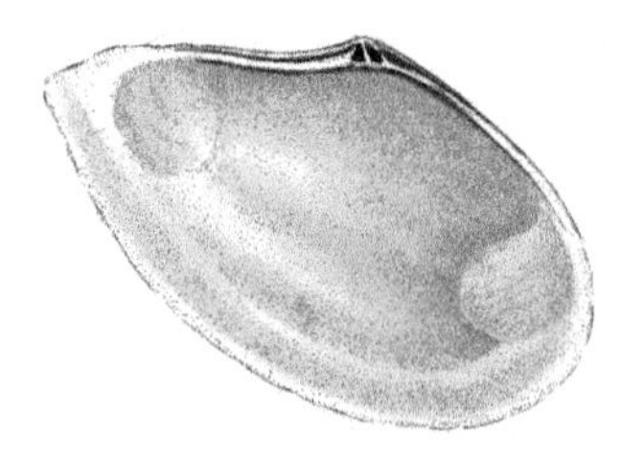

PLATE CLXIII.

TELLINA DEPRESSA.

DEPRESSED TELLEN.

GENERIC CHARACTER.

The hinge usually furnished with three teeth. Shell generally sloping on one side.

SPECIFIC CHARACTER

AND

SYNONYMS.

Shell inæquilateral, depressed, and very minutely striated.

TELLINA DEPRESSA: testa inæquilatera depressa minutissime striata.
Gmel. Linn. Syst. p. 3238. *sp.* 55.
Gualt. test. t. 88. *f. L.*

Tellina Squalida. *Soland Mus. Port.—Pult. Cat. Dors. p.* 29.

In the summer of the year 1800, we first discovered this shell, laying in plenty upon the sands on the south east side of Tenby, Pembrokeshire; where they had been apparently thrown up by a violent sea that had raged with considerable fury two or three hours before. This shell we conceived to be an undescribed British shell, but have since found that it had been observed on the north shore of Poole, at Weymouth, sparingly, by Dr. Pultney, and described by him in

PLATE CLXIII.

Hutchin's History of Dorsetshire, under the specific name of *Squalida.* This conchologist admit it to be Tellina squalida of Solander, *Mus. Port.* and Tellina depressa of Gmelin.

The only synonym given by Gmelin for his *T. depressa,* (whose *habitat* he is even unacquainted with,) is a reference to Gualtieri, f. H. I. L. The two first are small, and perhaps not of the same species, those represented at letter L and M, we believe to be the true shell of which we offer a figure as the Tellina depressa of Gmelin; Gualtieri thus describes his shell: "Tellina inæquilatera satis depressa, minutissime striata, vel candida, vel purpurascens, vel subrosea."

The figures in the annexed plate represent the natural size of our largest specimens: the colours are variable, more or less, of a fine pale orange, yellow, and tinged with rosy. It is certainly rare.

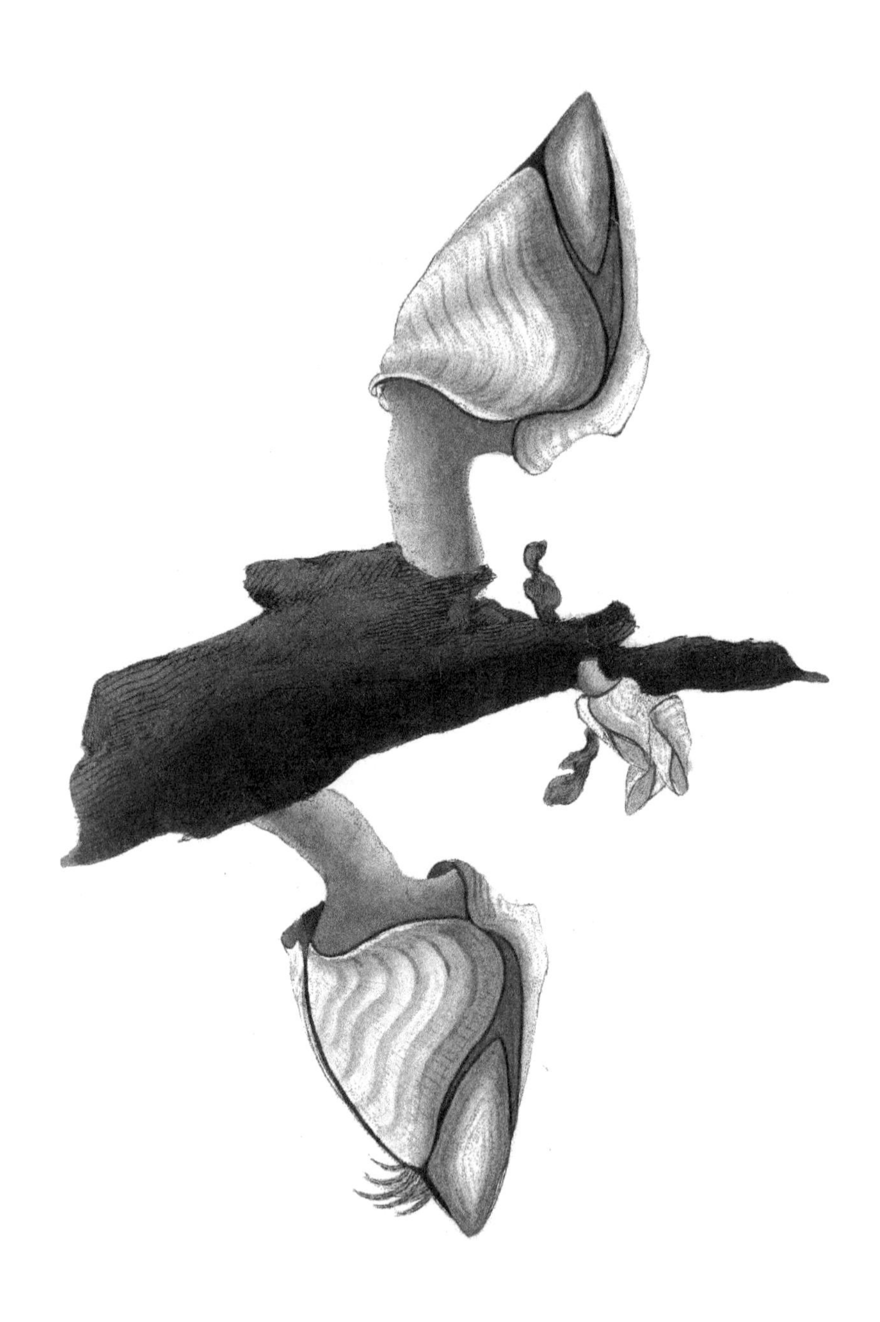

PLATE CLXIV.

LEPAS DILATA.

DILATED LEPAS.

GENERIC CHARACTER.

Animal Triton. Shell affixed at the base: multivalve; the valves unequal.

SPECIFIC CHARACTER
AND
SYNONYMS.

Shell compressed, five valved, thin, dorsal valve dilated at the base with an acute angle; and seated on a peduncle.

LEPAS DILATA: testa compressa quinquevalvi tenui, valvula dorsali basi dilata angulo acuto, pedunculo insidente.

LEPAS FASCICULARIS: testa quinquevalvi lævi corpus tegente, valvula dorsali basi dilata angulo acuto prominente, stipite nudo. *Ellis. Zooph. pl.* 15. *f.* 6. *p.* 167.

LEPAS SIGILLATUM, *Mus. Portl?*

The first, and only account we have of this kind of Lepas, is that given by the late Mr. Ellis in his Natural History of Zoophytes. In addition to the character he assigns to this shell, quoted as a synonym, this writer acquaints us only that it is "from St. George's Channel." We have never met with it either on the coast of that channel, or any other, nor have we received it from any of our friends, at the same time that we have no reason to distrust the information of that author, and only infer from the attention we have by chance bestowed particularly to the marine productions of that sea, that it must be rare.

A specimen of this shell, one which we are inclined to think, on pretty good authority, to be the same, or one of them at least, that was sent by the late Mr. Ellis to the Dutchess of Portland, is at this time in our Cabinet; the late Dr. Fordyce became first possessed of this specimen, and at his death we obtained it, under the title of Lepas sigillatum of Solander. Unlike Lepas anatifera, or anserifera, the valves of this shell are uncommonly thin, brittle, in a certain degree corneous, with the largest lateral valves rather crumpled in the usual course of the striæ, and marked transversely with obsolete rays: the shell is likewise covered with a fine pale brown skin, or epidermis: is larger than anatifera, and has a singular acute prominent dilation at the base of the back valve.

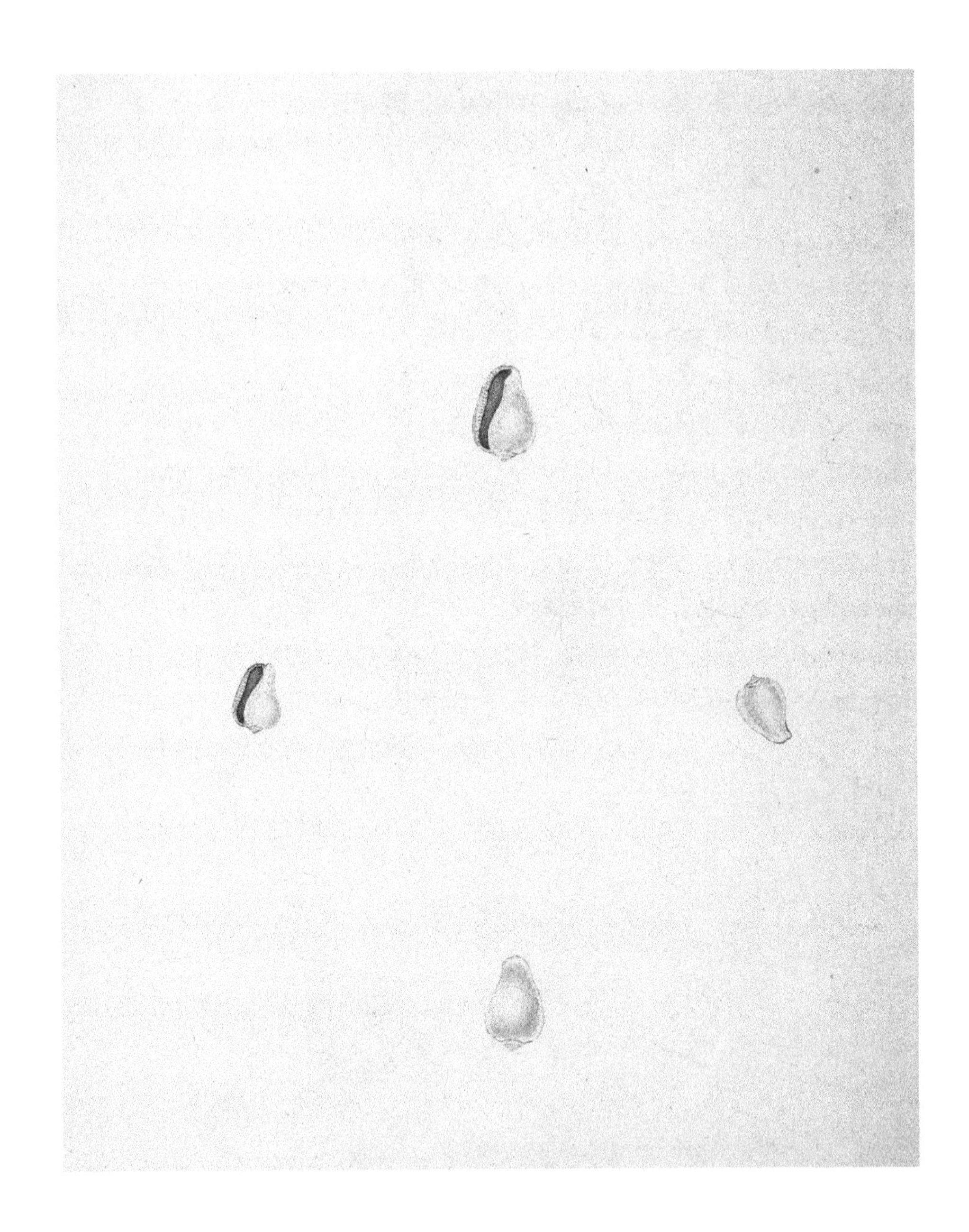

PLATE CLXV.

VOLUTA LÆVIS.

SMOOTH VOLUTE.

GENERIC CHARACTER.

Animal a Limax. Shell with one cell, spiral; aperture without a tail or beak, and somewhat effuse. Columella plaited; generally without lips or umbilicus.

SPECIFIC CHARACTER.

Shell rather ovate, very smooth; spire obtuse; two plaits on the pillar lip; lip gibbous, and slightly denticulated.

VOLUTA LÆVIS: testa obovata lævissima, spira obtusa, columella biplicata, labro gibbo subdenticulato.

VOLUTA EDENTULA. *Mus. Portl.*

As a British species, this extremely rare little shell was first noticed on the coast of Weymouth, being dredged up in deep water by some fishermen, and consigned to the cabinet of the late Dutchess of Portland. The specimens we have figured, are two of those originally in the possession of her Grace.

PLATE CLXV.

Dr. Solander, who, it is well known to the scientific conchologist, intended to have published a catalogue of that Museum, it appears, on a reference to his posthumous papers, called this species edentula ; a name which, without detracting from the merit of that able naturalist, it must be allowed is by no means applicable. So far from its being destitute of teeth, the series of denticulations are sufficiently visible on the *columella* ; those on the lip are yet more prominent, and can by no means justify the appellation of edentula.

Voluta Lævis, for such we have presumed to name this shell, is remarkably glossy, free in a perfect degree from any kind of striæ, whitish, and most delicately tinged with pale blushes of red, and yellowish or straw colour.—It has much the habit of a cypræa, and might without any impropriety be arranged under that genus.

1

2

PLATE CLXVI.

FIG. I.

LEPAS SCALPELLUM.

GENERIC CHARACTER.

Animal Triton. Shell affixed at the base, multivalve; the valves unequal.

SPECIFIC CHARACTER AND SYNONYMS.

Shell compressed. Valves thirteen, smooth, and seated on a scaly peduncle.

LEPAS SCALPELLUM: testa compressa tredecim valvi læviuscula pedunculo squamosa insidente. *Linn. Fn. Suec.* 2121.—*Gmel. Syst. Nat.* 3210. *Sp.* 11. *Ellis Phil. Trans.* 1758. *t.* 34. *f.* 4. *page* 849.

Lepas Scalpellum, a very rare and curious species, has been found attached to some sea weeds, dredged up on the coast of Weymouth; a specimen of it affixed to the branches of a coralline that was discovered here, after passing through the collections of the late Dutchess of Portland, and Dr. Fordyce, is at present in our possession.

There are several interesting remarks upon this singular genus in a paper written by the late Mr. John Ellis; which is inserted in the transactions of the Royal Society, for the year 1758: the letter is addressed to Mr. Isaac Romilly, a member of the society, and contains in particular, the following observation upon *Lepas Scalpellum.* "Fig. 2," he says, referring to his illustrative plate, "is the next animal of this class: this is not yet described. I found several of them sticking to the warted Norway Sea Fan, which Dr. Pantoppidan, the Bishop of North Bergen, sent you: from its appearance, I have called it the Norway Sea Fan Penknife. The stem of this is covered with little testaceous scales. The upper part of the animal is enclosed in thirteen distinct shells, six on each side, besides the hinge-shell, which is common to both sides: these are connected together by a membrane that lines the whole inside.

Gmelin speaks of it as a native of the Norway seas.

FIG. II.

LEPAS ANSERIFERA.

SPECIFIC CHARACTER AND SYNONYMS.

Shell compressed, quinquevalve, striated, and seated on a peduncle.

LEPAS ANSERIFERA: testa compressa quinquevalvi, striata, pedunculo insidente. *Gmel. Linn. Syst. p.* 3210. *Sp.* 12.

PLATE CLXVI.

Chiefly distinguished from Lepas Anatifera, described at the commencement of this work, by having the valves striated with elevated lines; the valves in the former being perfectly smooth. Lepas anatifera has been heretofore considered as a native of the American and Atlantic seas; but that it has been likewise found upon the English coast, there is no reason to dispute, the shell with the living animal has been dredged up at Weymouth, as well as the preceding species. We have the valves of this shell likewise in the collection of Da Costa, as an English species.

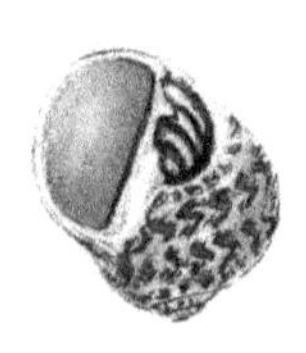

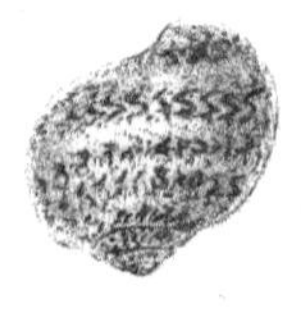

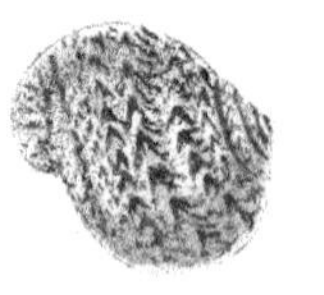

PLATE CLXVII.

NERITA INTRICATA.

INTRICATE-LINED NERIT.

GENERIC CHARACTER.

Animal Limax. Shell univalve, spiral, gibbous, flattish beneath: aperture semiorbicular, or semilunar; pillar lip, transversely truncated, and flattish.

SPECIFIC CHARACTER.

Shell smooth: spire somewhat pointed: umbilicus large, nearly heart shaped, with a small carinated lobe.

NERITA INTRICATA: testa lævi: spira submucronata, umbilico magno subcordato; lobo parvo carinato.

NERITA CANRENA *var?*

That this shell is not the young of Nerita Glaucina, as some have suspected, is evident from the depth and structure of the umbilicus, which in the former is almost completely closed by the pillar lip. It is rather allied to Nerita Canrena, which has a gibbous bifid umbilicus; and may possibly indeed prove to be nothing more than a variety of that shell. The varieties of N. Canrena, enumerated by Gmelin, amount to twenty-five, neither of which accords exactly with our shell, and that writer describes them only as natives of India, Africa, and America, but it is not unlikely it may be also an European shell.

Our specimens are from Weymouth.

1
2
2
1

PLATE CLXVIII.

FIG. I.

HELIX PUTRIS.

MUD SNAIL.

GENERIC CHARACTER.

Aperture of the mouth contracted and lunulated.

SPECIFIC CHARACTER

AND

SYNONYMS.

Shell imperforate, obtuse, ovate, yellow: aperture ovate.

HELIX PUTRIS: testa imperforata, ovata obtusa flava: apertura, ovata. *Linn. Fn. Suec.* 2189.
Gualt. t. 5. *f. H.*
Chem. Conch. 9. *t.* 135. *f.* 1248.
List. Conch. t. 123. *f.* 23 ?

G 2

PLATE CLXVIII.

Helix (succinea) testa oblonga fulva diaphana, anfractibus tribus, apertura ovata. *Mull. Zool. dan. prodr.* 2912.

Turbo subflavus pellucidus imperforata, testa prætenui fragili, trium spirarum. TRIANFRACTUS, THREE SPIRED. *Da Costa, p.* 92. *ſp.* 51. *Pl.* 5. *fig.* 13.

The two Linnean species of fresh water helices, *putris*, and *limosa* are so closely allied, that authors have, not unfrequently, confounded one with the other. Dr. Pultney considers our shell with some doubt, as the Helix limosa of Linnæus, in which particular we think, he must be mistaken. The figure given by Gualtieri is the only one referred to by Linnæus, in the Systema Naturæ, for Putris, and that is most exactly the same as our shell, although the design is taken from a minute specimen. Pennant's Helix Putris, t. 86. f. 137. is apparently the H. limosa. Both the species in question are well figured by Chemnitz, on the same plate.

This shell is very common in ditches, ponds, and other watery places, and especially in those overgrown with weeds.

FIG. II.

HELIX BULLAOIDES.

SPECIFIC CHARACTER.

Shell ovate, smooth, glossy, horny, brittle, whorls reversed, spire short.

PLATE CLXVIII.

HELIX BULLAOIDES: testa ovata lævi nitida cornea fragili, anfractibus contrariis, spira brevi.

Received from Lincolnshire by the late Duchess of Portland, from whose collection the specimen at this time in our possession was obtained.

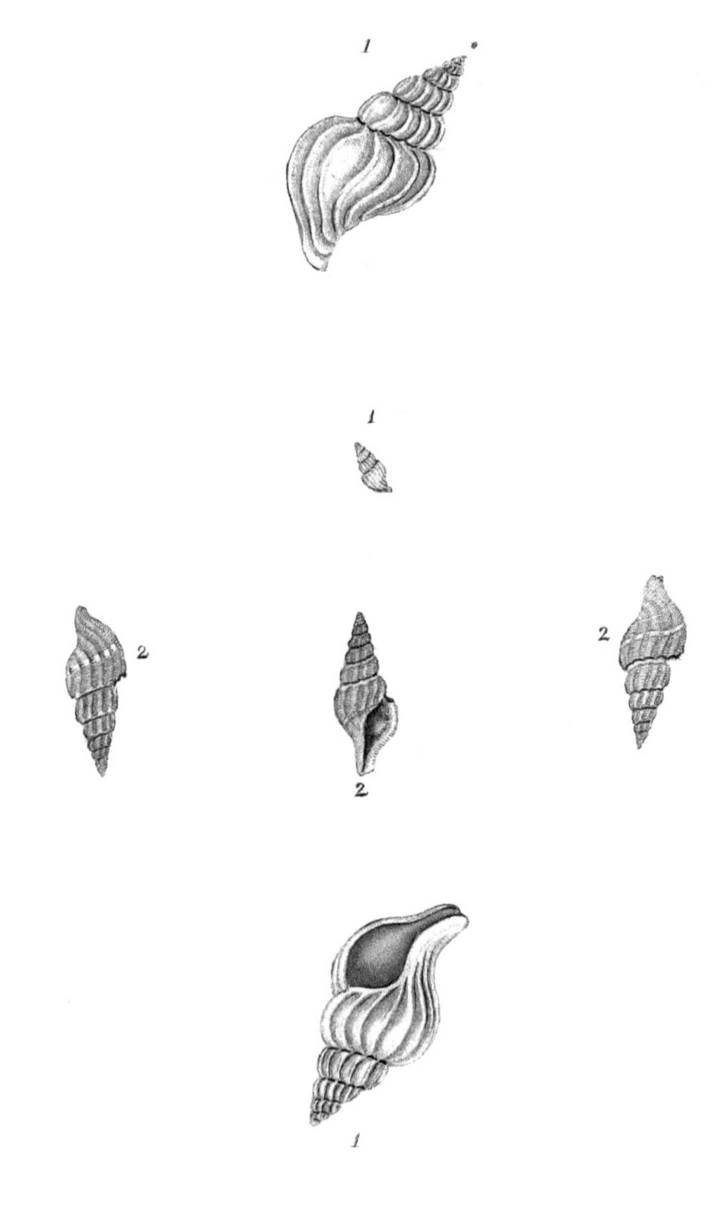
1
1
2
2
2
1

PLATE CLXIX.

FIG. I.

MUREX BAMFFIUS.

BAMFF WHITE MUREX.

GENERIC CHARACTER.

Spiral, rough. The aperture ending in a ſtrait, and ſomewhat produced gutter, or canaliculation.

SPECIFIC CHARACTER.

Shell ventricoſe, white, ribbed longitudinally, with acute plaits.

MUREX BAMFFIUS: testa ventricosa alba costis longitudinalibus acutis plicatis.

A nondescript species, discovered by Mr. Cordiner, on the coast of Bamffshire, Scotland; and communicated by him to the late Duchess of Portland. The smallest figure is only of the young shell, we have it of the exact size of the largest figure, numbered 1, in the annexed plate. Uncommonly rare.

G 4

PLATE CLXIX.

FIG. II.

MUREX EMARGINATUS.

NOTCHED-LIP PALE MUREX.

SPECIFIC CHARACTER.

Shell somewhat elongated, pale, with a white band: wreaths striated, with longitudinal undulations: on the posterior part of the lip a single notch.

MUREX EMARGINATUS: testa sub-elongata pallida fusca alba: anfractibus striatis longitudinaliter undulatis labio postice emarginato.

The notch in the posterior part of the lip of this shell is singular. By this mark the species may be immediately distinguished in a collection of British Shells, being perhaps the only one of the kind found on our coast. This notch, it should however be added, is to be considered rather as the distinctive feature of a natural family of shells, than as the character of the individual species now before us, the very same appearance being observable on several of the extra-european shells of the Murex Genus.

Our specimens of this scarce, and, as we believe, undescribed species, were found on the western coast.

170

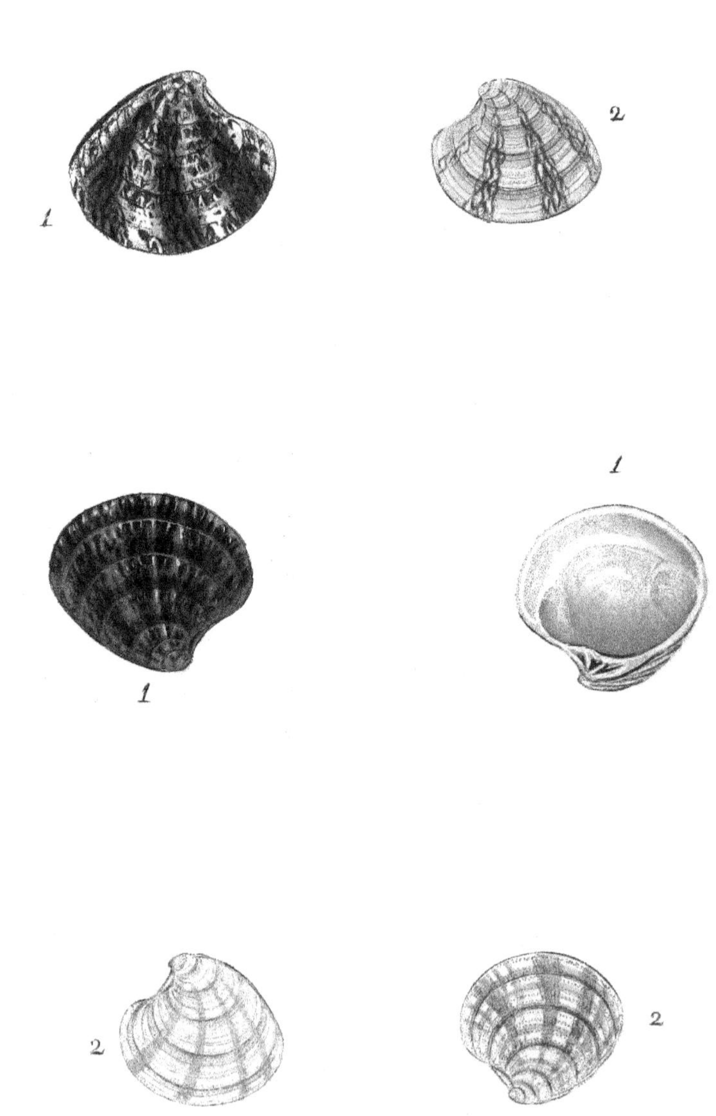

PLATE CLXX.

VENUS FASCIATA.

FASCIATED VENUS SHELL.

GENERIC CHARACTER.

Hinge furniſhed, with three teeth, two near each other, the third divergent from the beaks.

SPECIFIC CHARACTER

AND

SYNONYMS.

Shell somewhat heart-shaped, white, fasciated with brown: ridges large, broad, depressed, of regular thickness at both extremities.

VENUS FASCIATA: testa subcordata alba fusco-fasciata sulcis crassis latis depressis regularibus continuis.

Pectunculus parvus, planior, crassus dense fasciatus. Fasciated: *Da Costa, p.* 188. *sp.* 25. *Tab.* 13. *fig.* 3.

This elegant shell occurs very rarely on the coasts of this country; we have observed it sparingly distributed on the sands near Tenby, in Pembrokeshire. Da Costa says, he was informed that it is found near Bangor, among the rocks from Bangor Ferry to Anglesea, in

Wales, by which he could only mean that the species is an inhabitant of the Menai, the arm of Beaumaris bay, communicating with the St. George's channel which divides Caernarvonshire from the island of Anglesea. The same writer notes it likewise from Cornwall. Dr. Pultney describes it as a scarce shell, which he had found at Weymouth.

Having Da Costa's specimens of this shell, and also that of his Pectunculus Vetula before us, we should not refrain from observing, that the opinion of Dr. Pultney respecting these shells is incorrect; they are not merely transitions in growth, or varieties of the same kind, the difference between the two is obvious, and fully authorize us to consider them as distinct species. It should be understood in advancing this remark, that the shell which Da Costa figures and describes, for Pectunculus Vetula is clearly the Linnæan Venus Paphia, a shell well known as a native of the West Indies, and never found to our knowledge in any of the European seas. Da Costa was aware, after his work had been published, that he had erroneously confounded the variety of Fasciatus, Fig. 1, 1, in our Plate, with the West Indian shell; he had conceived the latter to be the same shell in a more perfect condition, and caused it to be engraved accordingly.

Dr. Pultney, in the passage wherein these shells of Da Costa are noticed (in his catalogue of the shells found on the coast of Dorsetshire,) describes the Pectunculus Fasciatus as nothing more than a variety of Venus Paphia (*Linn.*) in which respect he is assuredly mistaken. One of the most striking characters, by means of which the two species are to be discriminated, in our opinion, may be observed in the structure of the concentric ridges on the outside of the shell: these in the true Linnæan Paphia are remarkably thick, and prominent in the middle, but in approaching each extremity become suddenly obtuse, and are then continued in an attenuated ridge, particularly as they extend towards the front of the shell, and thus exactly corresponding with the definition of Linnæus, " rugis incrassatis, pube rugis attenuatis." On the contrary, in our shell the ridges are nearly of an uniform thickness throughout, sloping gradually with the depression of the shell behind, and only terminating abruptly at the edge of the front, or fore part of the shell where the valves appear obtuse: the outline of the shell is also very different from Venus Paphia, the latter being more produced on each side than our Venus Fasciata.

171

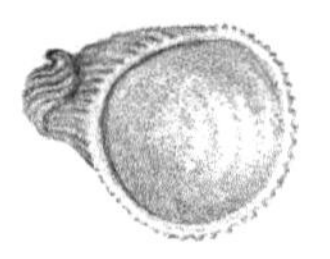

PLATE CLXXI.

PATELLA MILITARIS.

HOOKED LIMPET.

GENERIC CHARACTER.

Animal a Limax : shell univalve, sub-conic without spire.

SPECIFIC CHARACTER.

Shell entire, conic, pointed, striated, with the tip hooked, or recurved on one side.

PATELLA MILITARIS : testa integra conica acuminata striata, vertice hamoso lateraliter recurvo. *Linn. Mant.* 552.
List. Conch. 544.
Pult. Cat. p. 51.

When Linnæus described this shell in the Appendix to his *Mantissa Plantarum*, its native country was unknown to him. It is a Mediterranean species, and is sometimes, though rarely, met with on the British coast. We have it from Cornwall through the favour of Miss Pocock, and lately from Devonshire. Dr. Pultney acquaints us, that Mr. Bryer found this species on the sands near Weymouth Castle, Dorsetshire.

PLATE CLXXI.

Gmelin, in his edition of the *Systema Naturæ*, neglects to insert this species, for what reason we are at a loss to conceive. The specimens we possess of this rare shell, from the warmer parts of Europe, are larger than those found on our coast. In different specimens we observe that the striæ are liable to vary both in form and number, some shells appearing much more strongly reticulated than others.

172

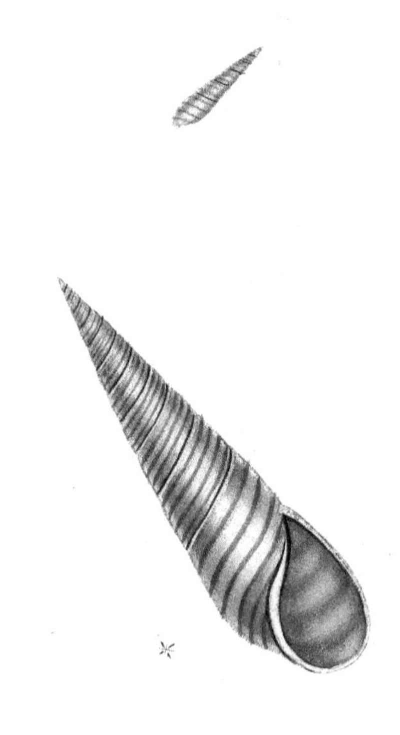

PLATE CLXXII.

TURBO SUBULATUS.

SUBULATE WREATH SHELL.

GENERIC CHARACTER.

Animal Limax. Shell univalve, spiral, or of a taper form. Aperture somewhat compressed, orbicular, entire.

SPECIFIC CHARACTER
AND
SYNONYMS.

Shell subulate, tapering, pale flesh-colour, glossy, fasciated with testaceous-brown. Aperture oval.

TURBO SUBULATUS: testa subulato-turrita pallide-carnea nitida testaceo fasciata, apertura ovali.

STROMBIFORMIS parvus corneus glaber. Smooth. *Da Costa, Brit. conch. p.* 117. *sp.* 69. Turbo lævis. Smooth. *Penn. Brit. Zool. No.* 115. *tab.* 79. *upper figure?*

Our best specimens of this rare shell were dredged up on the coast of Weymouth. Da Costa received it from Exmouth, in Devonshire; he also adds, that three were found in the stomach of a Five Finger, or common Stella Marina. The Turbo lævis of Pennant is from the coast of Anglesea, but it is altogether uncertain whether he means this species or not.—It is an elegant shell, of a taper form, thin, and semitransparent ; when very perfect, of a pale flesh-colour, spirally wreathed with whitish lines, and others of an ochreous or brownish hue ; the stripes are not uniformly disposed alike in all specimens. Da Costa thinks the species may be well distinguished by the spiral white lines.

The smallest figures in the plate denote the natural size of this shell.

PLATE CLXXIII.

TURBO MAMMILLATUS.

MAMMILLATED TURBO.

GENERIC CHARACTER.

Animal Limax: shell univalve, spiral, or of a taper form. Aperture somewhat compressed, orbicular, entire.

SPECIFIC CHARACTER.

Shell imperforate, subovate, whorls striated with raised dots, and slightly angulated by a few of the striæ, the dots of which are larger.

TURBO MAMMILLATUS: testa imperforata subovata anfractibus striatis punctis eminentibus striis aliquot majoribus subangulatus.

This remarkable shell is introduced among the British species of the Turbo genus, only on the authority of a posthumous memorandum in the hand writing of Da Costa, which we find in the collection of that Conchologist affixed to one of the specimens figured in the annexed Plate. From this it appears the shell had been picked up by Mr. Platt on the Scilly rocks, at the western extremity of Cornwall, and communicated by him to Da Costa.

174

PLATE CLXXIV.

MYA PICTORUM.

PAINTER'S MUSCLE.

GENERIC CHARACTER.

Animal Ascidia. Shell bivalve, gaping at one end. The hinge for the most part furnished with a thick, strong, broad tooth, not inserted into the opposite valve.

SPECIFIC CHARACTER
AND
SYNONYMS.

Shell oblong, rounded at both ends; a single crenulated tooth in one valve, and two in the other.

MYA PICTORUM: testa oblonga antice posticeque rotundata, cardinis dente primario crenulato, alterius duplicato.

MYA PICTORUM. *Linn. Fn. Suec.* 2129.

Mya pictorum, Moule des Peintres. *Chemn. Conch.* 6. *t.* 1. *f.* 6. *Belg.* verf houder.—*Dan.* Maler miegen,—Maler Skiael, *ibid.*

Two, or perhaps no less than three different kinds of the fresh water *Myæ* have been confounded with the M. pictorum of Linnæus, as it has been already intimated in the description of the *Mya*

H 2

Ovalis, Plate 89, of this work, a shell considered by Da Costa as the true M. pictorum. The present species, which we have little hesitation in believing to be the shell Linnæus means, is more of an oblong form than *M. ovalis*, rounded at both extremities, thin, semi-transparent, and of a pale colour, beneath the epidermis, which is of a faint green, or brownish hue ; within the shell is pearly.

The only synonyms we can venture to adopt with certainty, are those above quoted. There can be no doubt that the figure given by Chemnitz for the M. pictorum of the continental writers, is the same as that now before us, but the extensive list of references to other works, added by that writer, we are induced to reject as being at least in many respects ambiguous. Our specimens of this shell were obtained from Mr. George Humphrey, who assures us, that although he never had met with it himself alive in this country, he was told some years ago by the late —— Seymour, Esq. that this very species had been fished up in the river Stour.

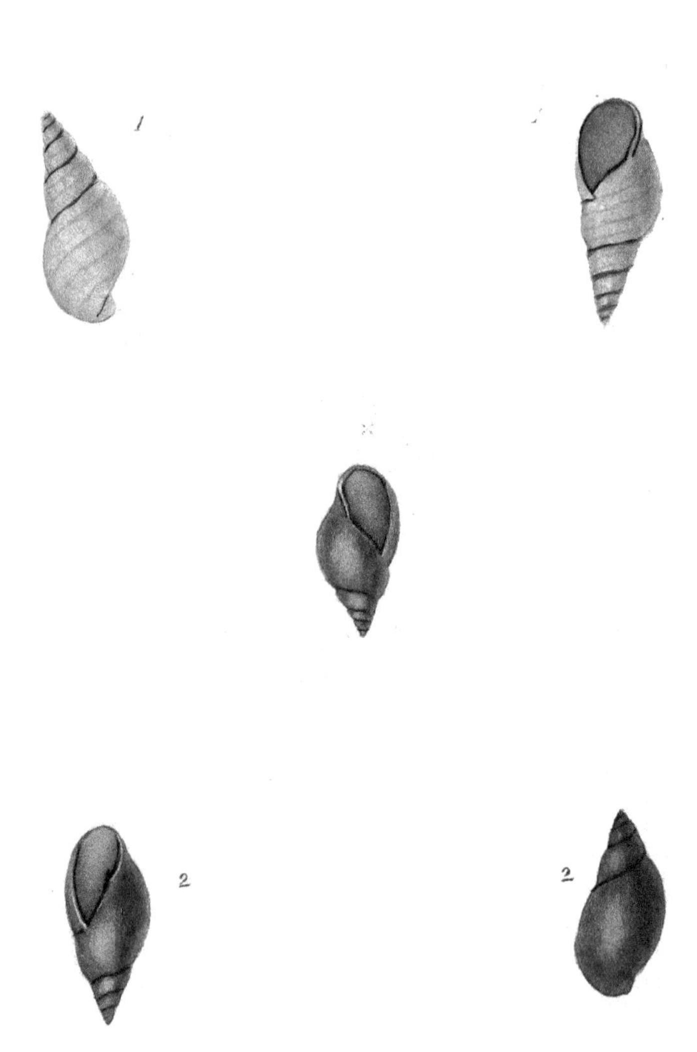
1
2
2

PLATE CLXXV.

FIG. I.

HELIX FRAGILIS.

BRITTLE RIVER SNAIL.

GENERIC CHARACTER.

Aperture of the mouth contracted, and lunulated.

SPECIFIC CHARACTER
AND
SYNONYMS.

Shell imperforate, ovate, tapering, round, pellucid: aperture oblong-ovate.

HELIX FRAGILIS: testa imperforata ovato-subulata tereti pellucida: apertura ovata-oblonga. *Linn. Faun. Suec.* 2187.—*Gmel. Linn. Syst. Nat.* 3658. *Sp.* 129. *Penn. Pl.* 86.

Brown River Snail. *Pult. Cat. p.* 48.

Helix fragilis is distinguished from Helix stagnalis, and one or two other very analogous species of river snail by a number of slight ridges which spirally traverse the whole shell, and are in particular

H 3 obvious

obvious on the firſt wreath. The shell is likewise more uniformly elongated than H. stagnalis, the first wreath being less swollen, or ventricose, and the remainder more so, than in that shell.—Helix fragilis we have found on plants growing in rivulets about Greenwich. Dr. Pultney says, it is common on plants in the river Stour.

FIG. II.

HELIX FONTINALIS.

SMOOTH FRESH WATER SNAIL.

SPECIFIC CHARACTER.

Shell imperforate, ovate, and pointed, glabrous, horny; volutions five, the first and second ventricose.

HELIX FONTINALIS: testa imperforata ovato-acuminata glabra cornea anfractibus quinque, primo secundoque ventricosis.

Common in some of the rivulets in Devonshire. Communicated by J. Laskey, Esq.

Fig. 2, are those of the common sort; a reversed variety of the same species is distinguished by a star.

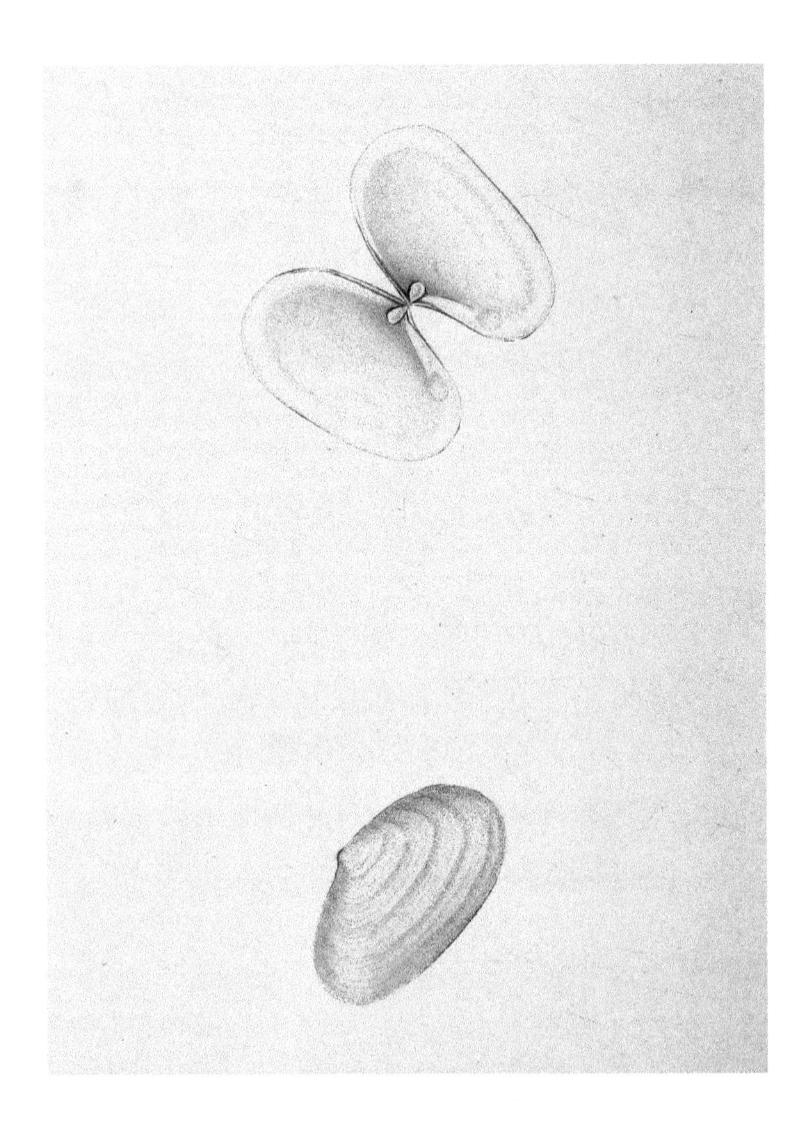

PLATE CLXXVI.

MYA PRÆTENUIS.

THIN WHITE SPOONHINGE GAPER.

GENERIC CHARACTER.

Animal an Ascidia. Shell bivalve, gaping at one end. The hinge for the most part furnished with a thick, strong, and broad tooth, not inserted into the opposite valve.

SPECIFIC CHARACTER

AND

SYNONYMS.

Shell ovate, subpellucid, fragile, white; tooth in both valves at the hinge oval, and patulous.

MYA PRÆTENUIS: testa ovata subpellucida fragili alba, cardinis dente ovali patulo.

MYA PRÆTENUIS: testa ovata subpellucida fragili alba subumbonali pubescente, cardinis dente ovali patulo. *Pult. Cat. p.* 28.

CHAMA *prætenuis* cardine cochleato porrectiore. *Petiv. Gaz. t.* 94. 4.

This delicate shell was first observed by us on the sandy coast of Caermarthenshire. We have since received the same kind from the shores of Cornwall, where it was found by Miss Pocock. Petiver

H 4

describes his shell as being found at Poole, in Dorsetshire: where Dr. Pultney also met with it on the sands in the harbour; likewise on the north shore near Brownsea Isle, and once with a few valves on the shore between Weymouth and Portland.

Mya prætenuis, as the specific name implies, is a remarkably thin shell, very brittle, of a whitish colour, and distinguished by having an oval process or tooth resembling the bowl of a spoon in each valve at the hinge.

177

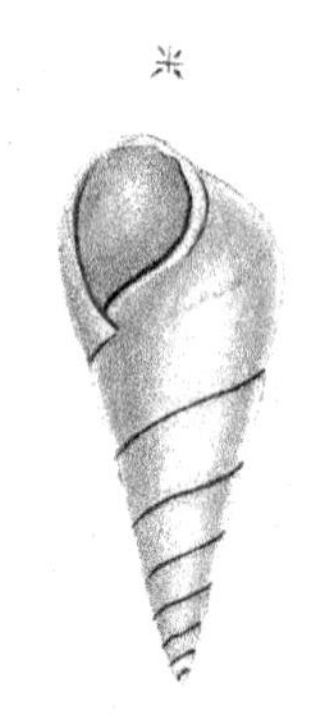

PLATE CLXXVII.

TURBO ALBUS.

WHITE WREATH SHELL.

GENERIC CHARACTER.

Animal Limax. Shell univalve, spiral, or of a taper form. Aperture somewhat compressed, orbicular, entire.

SPECIFIC CHARACTER AND SYNONYMS.

Shell tapering, glossy, and white.

TURBO ALBUS: testa turrita nitida alba.

Strombiformis parvus albissimus lævis, white. *Da Costa Brit. Conch. p.* 116. *Sp.* 68.

Turbo minimus lævis albus. Milk white smooth whelke. *Borlase Cornw. p.* 277.

Turbo albus. *Penn. Brit. Zool.* N° 114, *tab.* 79?

We have never met with this shell on any of the British sea coasts, although we are told it is found not very unfrequently on several of those to which our researches have been directed. Da Costa says, the species is found on the shores of Cornwall, about Fowey, Whitsand Bay, the Land's End, &c. and also in Devonshire. Pennant's Turbo Albus is from Anglesea.

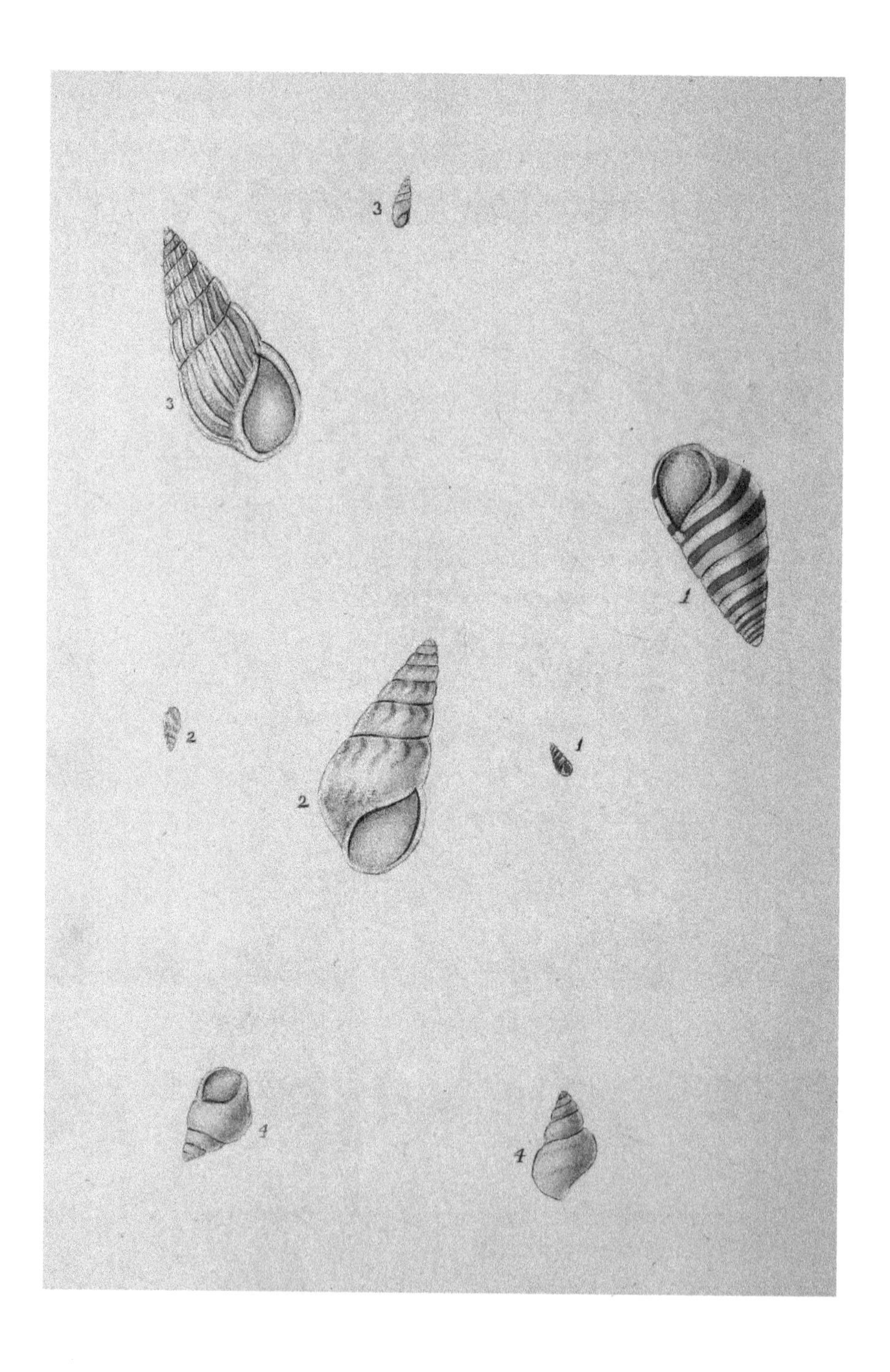
3
3
1
2
2
1
4
4

PLATE CLXXVIII.

FIG. I.

TURBO VITTATUS.

RIBBON-WREATH SHELL.

GENERIC CHARACTER.

Animal Limax. Shell univalve, spiral, or of a taper form. Aperture somewhat compressed, orbicular, entire.

SPECIFIC CHARACTER
AND
SYNONYMS.

Shell smooth, taper, whitish, whorls subobsolete ; on the first, three chesnut bands, one on the rest

TURBO VITTATUS : testa turrita albida anfractibus subobsoletis, primo faciis tribus castaneis reliquis unica.

This, and the following species of Turbo, we discovered in the Menai, between Caernarvonshire and the island of Anglesea. Turbo Vittatus, we have likewise been favoured with from Cornwall, by Miss Pocock, and from Devonshire, by J. Laskey, Esq.

PLATE CLXXVIII.

There is some reason for believing this to be the Turbo Trifasciatus of Adams's description of minute British Shells, discovered on the coast of Tenby, South Wales, which is inserted in the fifth volume of the Transactions of the Linnæan Society. The account he gives does not exactly agree with our Shell; he speaks of only two red bands on the first spire, instead of three; and the single spiral line arising from the posterior band, terminates in his Shell after encircling the second volution, whereas, in all our specimens this line is continued on every wreath to the apex. Should his *T. trifasciatus* be intended for our Shell, the outline also is very badly expressed.—The smallest figure in the annexed plate ſhews the natural ſize of this ſhell.

FIG. II.

TURBO INTERRUPTUS.

INTERRUPTED-STRIPE WREATH SHELL.

SPECIFIC CHARACTER

AND

SYNONYMS.

Shell smooth, taper, whitiſh, fasciated, with an interrupted ochreous band.

Turbo Interruptus: testa lævi turrita albida fascia interrupta ochrea.

Turbo Interruptus: teſta quinque anfractibus subobtusis, apertura subrotunda. *Adams. Linn. Trans. V. 5. ſp.* 3 ?

PLATE CLXXVIII.

We feel a much slighter degree of hesitation in admitting this to be the Shell meant by Mr. Adams, in the paper above quoted, than the preceding species; notwithstanding that our Shell has a greater number of whorls; the outline of his figure is rude, and far from characteristic of the shell.—Our specimens are from Anglesea, as before mentioned. It is represented both of the natural fize and magnified, in the plate.

FIG. III.

TURBO COSTATUS.

RIBBED WREATH SHELL.

SPECIFIC CHARACTER.

Shell taper, snowy white, with numerous obtuse longitudinal ribs.

TURBO COSTATUS: testa turrita nivea costis longitudinalibus numerosis obtusis.

Specimens of this elegant shell were found at Margate. The smallest figure denotes the natural fize.

FIG. IV.

TURBO PALLIDUS.

PALE WREATH SHELL.

SPECIFIC CHARACTER.

Shell somewhat taper, pale; whorls very slightly bicarinated.

PLATE CLXXVIII.

TURBO PALLIDUS: testa subturrita pallida anfractibus obsoletissime bicarinatus.

Found on the weſtern coast: a Shell of very plain appearance, brownish colour, and rather flattened on the wreath, so as to form two ſlight spiral ridges or obtuse angles, especially on the first or largest volution.

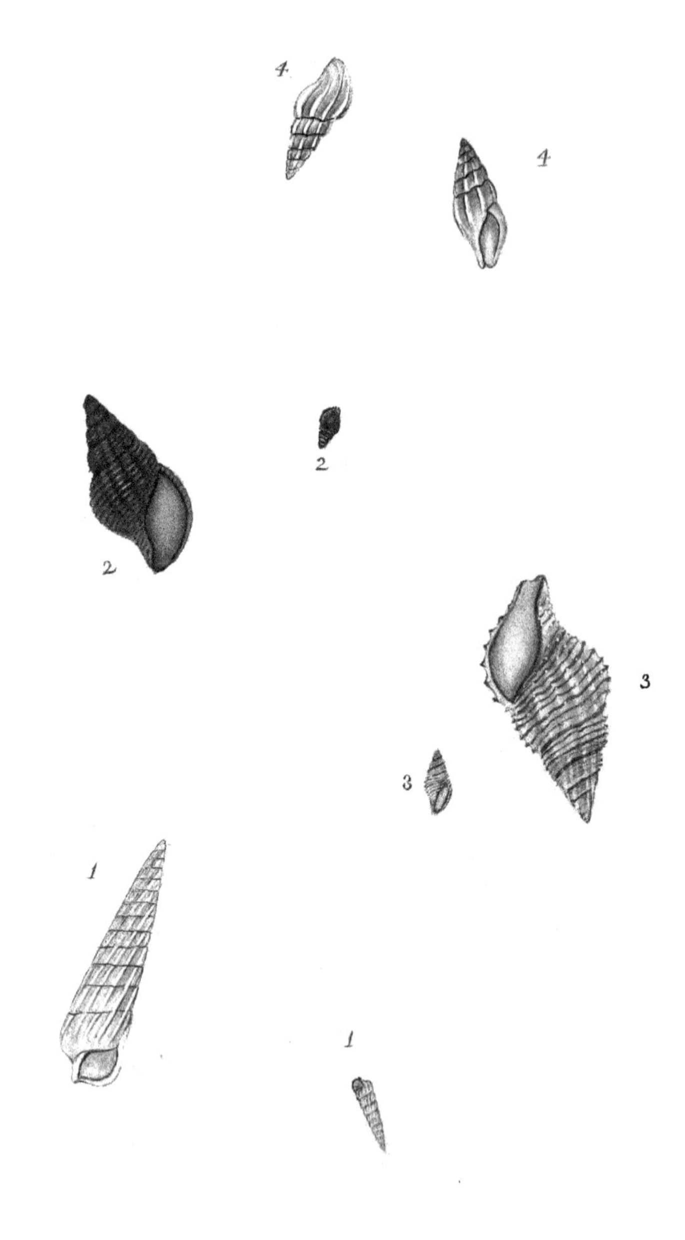
4
4
2
2
3
3
1
1

PLATE CLXXIX.

FIG. I.

TURBO ACUTUS.

ACUTE WREATH SHELL.

GENERIC CHARACTER.

Animal Limax. Shell univalve, spiral, or of a taper form. Aperture somewhat compressed, orbicular, entire.

SPECIFIC CHARACTER.

Shell taper, acute, snowy white, whorls about twelve with numerous oblique obtuse ribs.

TURBO ACUTUS: testa turrita acuta nivea, anfractibus subduodecim costis confertis obliquis obtusis.

A mutilated specimen of this curious Shell has been sent to us from the coast of Cornwall: we believe it has been also found at Weymouth; but our perfect shell of this species is from Guernsey.—The smallest figure is of the natural size.

PLATE CLXXIX.

FIG. II.

BUCCINUM BRUNNEUM.

LITTLE BROWN WHELK.

GENERIC CHARACTER.

Aperture oval, ending in a short canal.

SPECIFIC CHARACTER.

Shell taper, brown, whorls transversely striated, and longitudinally undulated. Aperture toothless.

BUCCINUM BRUNNEUM: testa turrita brunnea anfractibus transversim striatis longitudinaliter undulatis apertura edentula.

Found on the coast of Cornwall:—A rare, and, as it is presumed, an undescribed species.

FIG. III.

MUREX ELEGANS.

ELEGANT MUREX.

GENERIC CHARACTER.

Spiral, rough, aperture ending in a strait, and somewhat produced gutter or canaliculation.

PLATE CLXXIX.

SPECIFIC CHARACTER.

Shell yellowish, banded with ochreous, tip violet: whorls longitudinally ribbed, and finely striated transversely.

MUREX ELEGANS: testa ovata flavescente ochreo-fasciata apice violacea anfractibus longitudinaliter costatis transversim minutissime striatis.

A very beautiful little shell, discovered by Miss Pocock on the coast of Cornwall.

FIG. IV.

MUREX SEPTEM-ANGULATUS.

SEVEN-ANGLED.

SPECIFIC CHARACTER.

Shell oblong, acute, pale, with seven longitudinal angles.

MUREX SEPTEM-ANGULATUS: testa oblonga acuta pallida longitudinaliter septem-angulata.

This kind rarely occurs on our coasts. The specimens in our possession are from Weymouth. It bears a strong affinity to Murex costatus, plate XCIV. of this work, although it is certainly distinct. Murex costatus is much more linear in the outline, and has the ribs less prominent and acute than our Murex septem-angulatus.

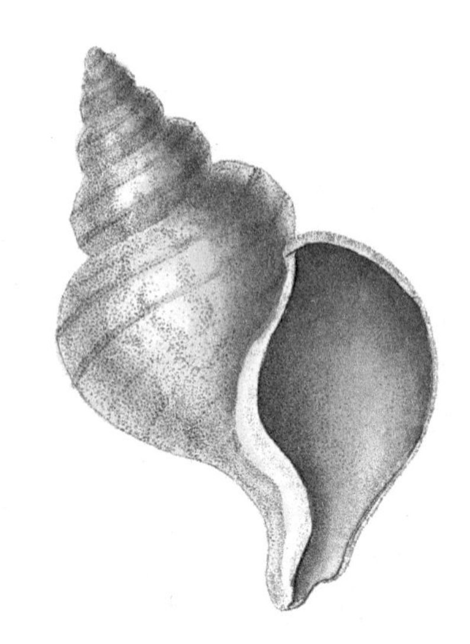

PLATE CLXXX.

MUREX DESPECTUS.

DESPISED MUREX.

GENERIC CHARACTER.

Spiral, rough. The aperture ending in a strait, and somewhat produced gutter, or canaliculation.

SPECIFIC CHARACTER
AND
SYNONYMS.

Tail patulous: shell oblong; whorls eight, with two elevated lines.

MUREX DESPECTUS: testa patulo-subcaudata oblonga: anfractibus octo lineis duabus elevatis. *Linn. It. Wgoth.* 200. *tab.* 5. *f.* 8.

The present shell, it must be tacitly acknowledged, is inserted among the rarer shells of this country on very slight authority; namely, that of a friend, who believes he once saw a few specimens of this *Murex* that were fished up in the sea at a short distance to the north of the Orknies.—On this suspicion only we

I 2

could not have presumed to insert the species in this work, were it not to avail ourselves of the opportunity afforded by that means to correct an error very generally admitted concerning the true Murex despectus of Linnæus, the shell at this time under consideration.

To the English conchologist it need be scarcely said, that another shell, somewhat similar to the present, although specifically different, has been hitherto received as the Murex despectus of Linnæus by every writer in this country who has had occasion to speak of that shell. The origin of this mistake, it will be perceived from the following particulars, rests in a great measure, if not entirely, with Linnæus himself. The Murex despectus of this writer is noticed, for the first time, in the account of his Travels through part of Sweden: a small octavo volume written in the Swedish language, with notes, relative to Natural History in Latin. At page 200, he describes this shell in these words, "cochlea spiris octo oblonga utrinque producta lineis duabus elevatis," referring to plate 8. fig. 5. of the same work for a delineation of the shell; the figure quoted in every respect agrees with our specimen, not only in the general outline, but most exactly in having the slight carinated ridges that pass spirally round the whorls, a character not observable on the Murex despectus of English authors. So far therefore we are convinced that the present shell is the Murex despectus of the Linnæan *Iter Westrogothicum.*

The work above mentioned appeared in 1746, the year in which Linnæus likewise published the first edition of his *Fauna Suecica.* In the latter, Murex despectus is again described with a reference to his *Iter W. goth.* and in addition to that synonym, a shell figured by Lister is also quoted for the same species. This is the source of that very confusion which has since arisen concerning the Linnæan

Despectus, and should be fully stated.—*Lister's Angl. t.* 3. *f.* 1. is the reference given by Linnæus Adverting to this we find the following definition of the shell given by Lister, " Buccinum album læve maximum septem spirarum."—He further adds, in the general description, " Testæ pars exterior ex tota lævis est, i. e. sine striis quamvis sæpius vel rugis quibusdam vel aliis rebus extrinsecus adnatis exasperetur." From this account, and from the figure he has given of the shell, there is not the smallest reason to dispute that Lister means the shell which English writers have heretofore considered as the Murex Despectus *; but it is not less certain that Linnæus was wrong in quoting Lister's figure for his Swedish shell, since they are not the same. However, on the authority of this reference to Lister, which afterwards appeared in the Systema Naturæ, this shell has continnued to be considered as the species meant by Linnæus.

Nor was this the only oversight which appears to have been committed by that eminent Naturalist; by continuing to refer, in the Systema Naturæ, to Lister's figure for his species Despectus, no one scarcely could imagine that Lister's shell should be the M. Antiquus of Linnæus, instead of his Despectus, and yet we are persuaded, after attentively comparing his description of the shells with his synonyms, that such is the fact : the description agrees with it, and the figure given by *Gualteri* is surely of the same kind as that which Lister speaks of.

The Linnæan shell, M. Despectus, is well described, and the figure in his *Iter. IV. Goth.* is expressive: the two elevated spiral lines, together with the rotundity of the wreaths, are strikingly

* In Lister's Plate the shell is reversed by mistake, most likely, of the engraver.

characteristic of this species. At the first glance this shell appears to be an intermediate kind between Lister's shell and the Murex Carinatus of Pennant, and ourselves: indeed the principal difference we perceive between the true M. Despectus and Lister's shell is, that the former has the whorls of the spire rather more ventricose, and distinctly marked with two slightly elevated spiral lines; from Murex Carinatus it differs principally in the very prominent angulations of the *anfractibus,* where the ridges appear, and more particularly in the strong depression between the upper ridge, and the suture of the whorls.

The Murex despectus, at pesent under consideration, is certainly very rare, except in the North of Europe, where we are led to suppose, from what Linnæus says, it is not uncommon.

The only specimens we have ever seen of this kind are from Greenland.

HAVING thus ascertained, as we may reasonably believe, the true Linnæan Murex Despectus, it remains in this place to propose the following emendations and additions to the description of two Shells figured in the course of this work, namely, Murex Despectus, Plate XXXI. and Antiquus, Plate CXIX. which, in common with other testaceological writers, we had misconceived.

PLATE XIX.

MUREX DESPECTUS,

read

MUREX ANTIQUUS.

ANTIQUATED MUREX.

Testa patulo-caudata oblonga: anfractibus octo teretibus. *Linn. Fn. Suec.* 2165.

PLATE CXIX.

MUREX ANTIQUUS,

read

MUREX DUPLICATUS.

TUBERCULATED MUREX.

Dele reference to *Linn. Fn. Suec. et Gmel. Syst. Nat.*

and

add for the

SPECIFIC CHARACTER

Shell patulous, tailed, oblong: whorls eight, tuberculated, striated, with two raised ridges.

MUREX DUPLICATUS: testa patulo caudata oblonga: anfractibus octo striatis duplicato carinatis: carinis tuberculatis.

I 4

INDEX

VOL. V.

LINNÆAN ARRANGEMENT.

MULTIVALVIA.

BIVALVIA. CONCHÆ.

INDEX.

INDEX TO VOL. V.

ACCORDING TO THE

HISTORIA NATURALIS TESTACEORUM BRITANNIÆ OF DA COSTA.

PART I.

UNIVALVA NON TURBINATA.

GENUS 1. PATELLA. LIMPET, FLITHER, OR PAP SHELL.

GENUS 3. SERPULA. THE WORM SHELL.

INDEX.

PART III.

UNIVALVIA TURBINATA.

GENUS 7. TROCHUS. THE TOP.

MARINÆ. SEA.

GENUS 9. HELIX.

* TERRESTRES. LAND.

** FLUVIATILES. RIVER.

GENUS 12. STROMBIFORMIS. NEEDLE SNAIL.

I N D E X.

ORDER 2.

BIVALVES.

GENUS 7. PECTUNCULUS. COCKLE.

MULTIVALVES,

GENUS 17. BALANUS. AÇORN.

ALPHABETICAL INDEX TO VOL. V.

INDEX.

FINIS.

Printed by Bye and Law, St. John's Square, Clerkenwell.

Lately Published,

BY THE SAME AUTHOR:

1. THE NATURAL HISTORY of BRITISH INSECTS; explaining them in their several States, with the Periods of their Transformations, their Food, Œconomy, &c. The whole illustrated with coloured Figures, designed and executed from living Specimens. In Ten Volumes, Royal Octavo, containing Three Hundred and Sixty Plates.—Price in Boards, 15l. 10s.

2. THE NATURAL HISTORY of BRITISH BIRDS; Or, a SELECTION of the MOST RARE, BEAUTIFUL, AND INTERESTING BIRDS which inhabit this Country. Embellished with One Hundred and Twenty-four Figures, drawn, engraved, and coloured from the original Specimens. In Five Volumes, Royal Octavo.—Price in boards, 9l.

3. THE NATURAL HISTORY of BRITISH FISHES, including scientific and general Descriptions of the most interesting Species, and an extensive Selection of accurately Finished Coloured Plates.—Taken entirely from Original Drawings, purpofely made from the Specimens in a recent State, and for the most part whilst living.

N. B. This Work is now in a Courfe of Publication, in Numbers, monthly, at 3s. 6d. each; and in Volumes, annually, at 2l. 2sl. each in Boards. It is intended to be comprised in Five Volumes, Royal Octavo. Two of these are completed, and may be had in boards, Price 4l. 4s.